COORDINATION POLYMERIZATION

Academic Press Rapid Manuscript Reproduction

COORDINATION POLYMERIZATION

A Memorial to Karl Ziegler

Edited by

JAMES C.W. CHIEN

Department of Chemistry
Department of Polymer Science and Engineering
Polymer Research Institute
Material Research Laboratories
University of Massachusetts
Amherst, Massachusetts

Academic Press, Inc. New York   San Francisco   London   1975
A Subsidiary of Harcourt Brace Jovanovich, Publishers

ACADEMIC PRESS, INC.
111 Fifth Avenue, New York, New York 10003

United Kingdom Edition published by
ACADEMIC PRESS, INC. (LONDON) LTD.
24/28 Oval Road, London NW1

Library of Congress Cataloging in Publication Data
Main entry under title:

Coordination polymerization.

Proceedings of a symposium sponsored by the American
Chemical Society and held April 3, 1974, at UCLA.
Includes bibliographies and index.
1.   Polymers and polymerization—Congresses.
2.   Coordination compounds—Congresses.   3.   Ziegler,
Karl Waldemar, (date)   —Congresses.   I.   Ziegler,
Karl Waldemar, (date)   II.   Chien, James C. W.,
(date)   III.   American Chemical Society.
QD380.C66   547'.28   75-33101
ISBN 0—12—172450—6

KARL ZIEGLER

CONTENTS

List of Contributors

D.G.H. Ballard, ICI Corporate Laboratory, The Heath, Runcorn, Cheshire, England

C. Carlini, Istituto di Chimia Organica Industriale, Universite di Pisa, 35, via Risorgimento, 56100 Pisa, Italy

E. Chiellini, Istituto di Chimia Organica Industriale, Universite de Pisa, 35, via Risorgimento, 56100 Pisa, Italy

J.C.W. Chien, Department of Chemistry, Department of Polymer Science and Engineering, Materials Research Laboratory, University of Massachusetts, Amherst, Massachusetts 01002

F. Ciardelli, Istituto di Chimia Organica Industriale, Universite di Pisa, 35, via Risorgimento, 56100 Pisa, Italy

F.S. Dyachovskii, Institute of Chemical Physics, Academy of Science USSR, Cherrogolovka, Norginsk District, Moscow 142432, U.S.S.R.

S. Fuji, Research Center, Mitsui Petrochemical Industries, Ltd., Waki-Mura, Kuga-Gun, Yamaguchi-Ken, 740, Japan

G. Henrici-Olivé, Monsanto Research S. A., Eggbühl-Strass 36 CH-8050 Zürich, Switzerland

J.T.T. Hsieh, Department of Chemistry, Department of Polymer Science and Engineering, Materials Research Laboratory, University of Massachusetts, Amherst, Massachusetts 01002

M. Julemont, Laboratoire de Chimie Macromoléculaire et de Catalyse Organique, Université de Liège, Sart-Tilman, 4000 Liège, Belgium

K.A. Jung, Department of Chemistry, University of Mainz, 65 Mainz, G.F.R.

T. Keii, Department of Chemical Engineering, Tokyo Institute of Technology, Meguro, Tokyo, Japan

W. Kern, Department of Chemistry, University of Mainz, 65 Mainz, G.F.R.

LIST OF CONTRIBUTORS

S. Olivé, Monsanto Research S. A., Eggbühl-Strasse 36 CH-8050 Zürich, Switzerland

A. Oschwald, Eidgenössische Technische Hochschule Zürich, Universität-strasse 6, CH-8006 Zürich, Switzerland

P. Pino, Eidgenössische Technische Hochschule Zürich, Universität-strass 6, CH8006 Zürich, Switzerland

H. Schenecko, Dunlop-Forschungslaboratorium, 645 Hanau, Postfach 129 G.F.R.

P.J.T. Tait, Department of Chemistry, University of Manchester Institute of Science and Technology, Manchester M61 1QD, England

Ph. Teyssié, Laboratoire de Chimie Macromoléculaire et de Catalyse Organique, Université de Liège, Sart-Tilman, 4000 Liège, Belgium

J.M. Thomassin, Laboratoire de Chimie Macromoléculaire et de Catalyse Organique, Université de Liège, Sart-Tilman, 4000 Liège, Belgium

E. Walckiers, Laboratoire de Chimie Macromoléculaire et de Catalyse Organique, Université de Liège, Sart-Tilman, 4000 Liège, Belgium

R. Warin, Laboratoire de Chimie Macromoléculaire et de Catalyse Organique, Université de Liège, Belgium

G. Wilke, Max-Planck-Institut fur Kohlenforschung, 433 Mülheim A. D. Ruhr, Kaiser-Wilheim-Platz, G.F.R.

Yu. I. Yermakov, Institute of Catalysis, 630090 Novosibirsk, U.S.S.R.

V.A. Zakharov, Institute of Catalysis, 630090 Novosibirsk, U.S.S.R.

A. Zambelli, Instituto di Chimica delle Macromolecole, via Alfonso Corti N. 12, 20133 Milano, Italy

PREFACE

It was exactly 20 years ago that Ziegler discovered the catalyst system that made it possible to polymerize ethylene at low pressure to linear high density polyethylene. Subsequently Natta developed catalyst systems for stereospecific polymerizations of α-olefins. Soon it was realized that the mechanism of these polymerizations is distinctly different from others and that the propagating chain is coordinated to the catalyst. The principle of coordination polymerization has been applied successfully to other monomers. The Polymer Chemistry Division asked M. Tsutsui and myself to organize a symposium to look at what lies ahead for coodination polymerization.

While we were organizing this symposium, the news of the death of Professor Ziegler shocked the scientific community. The organizers proposed to the Polymer Chemistry Division and received unanimous approval to make the symposium a memorial to K. Ziegler. We were able to bring together for the occasion many scientists who are major contributors to the field of Ziegler-Natta catalysis. Because of the limitation of time it was possible to accommodate only a limited number of speakers. Other scientists were invited to contribute additional papers. This book contains all these contributions.

In Chapter 1, Wilke, who succeeded Ziegler as Director of the Max-Planck-Institut für Kohlenforschung, traces Ziegler's various major contributions to the chemistry of free radicals, organo-alkali metal compounds, many membered rings and organotransition metallic compounds.

In Chapters 2 and 3 the origin of steric control in the polymerization of α-olefins is discussed. In particular the elegant work of Pino and colleagues on steroselection and stereoelection shows that it is the chirality of the transition metal which dictates the stereoregularity of the polymer produced.

The counting of active sites has been a problem of central importance to the mechanism and kinetics of coordination polymerizations. Chapters 4–7 deal with this problem in depth and show comparisons of results obtained with various methods. These chapters are particularly useful because the limitations in each method are clearly spelled out for the benefit of others who might wish to employ the techniques. With the concentrations of active sites thus determined, some absolute rate constants are presented in these chapters.

To elucidate the mechanisms of initiation, propagation, termination and transfers. Chapter 8 deals with polymerizations by $(\pi\text{-}C_5H_5)_2TiCl_2$ catalysts. Chapter 9 describes the polymerizations by allyl, benzyl, trimethylsilmethyl and other derivatives of Ti and Zr. The kinetics of Ziegler-Natta polymerization is treated, along with the Langmuir adsorption mechanism, in Chapter 10.

The real new areas of Ziegler–Natta polymerizations are discussed in the last two chapters. In Chapter 12 results of supported catalysts are presented. The factors influencing the stereoregularity of polymerization of propylene by supported catalysts are investigated and delineated. Results of ethylene polymerizations by supported catalyst are also contained in Chapters 5 and 9. In Chapter 13 the interesting polymerizations of diolefins to equibinary polymers by h^3-allylic coordination complexes are discussed.

The organizers wish to acknowledge financial contributions from Dow Chemicals, E. I. duPont de Nemours & Co., Witco Chemical Corporation, Farbwerke Hoechst AG, and the Polymer and Science Engineering Department of the University of Massachusetts. These funds made it possible to bring the foreign scientists to the Symposium.

Finally, Frau Maria Ziegler was kind enough to send us her favorite photo of Professor Ziegler for this memorial publication. The editor is also grateful to the National Science Foundation for support through grant GH-39111.

James C. W. Chien

Karl Ziegler, in Memoriam

G. WILKE

Max-Planck-Institut für Kohlenforschung
Mülheim a.d. Ruhr
Germany

Karl Ziegler died in Mülheim a.d. Ruhr on the 11th
August 1973 shortly before his 75th birthday. A week later
we heard that the ACS had decided to hold a Memorial Symposium in his honor as part of the spring meeting in Los
Angeles. The news of this decision coming as it did so soon
after his death made a great impression upon Frau Ziegler and
the close friends and relatives and Frau Ziegler has asked me
to convey her thanks to the organizers of the ACS for this
gesture.
Remarkable in Karl Ziegler's career is the early date at
which he started in the direction which many years later was
to be crowned with such considerable success. Ziegler was
born on the 26th November 1898 in Helsa near Kassel the
son of a minister (who had studied theology at the University
of Marburg) and his interest in chemistry stems from a home-
laboratory in which he was working already at the age of
eleven. By the time he went to the University of Marburg to
study chemistry; he had obtained such a wide knowledge that
after three or four weeks he was allowed to transfer directly
to the third semester. For this reason it is not surprising
that he received his doctoral degree shortly before his 22nd
birthday. His Doktorvater was Karl von Auwers who very soon
encouraged him to make himself independent. Ziegler married
in 1922 and submitted his "Habilitations" thesis in 1923.
This thesis contained the first studies of the addition of a
metal-carbon bond to a C=C double bond which was later to
become such an important and versatile reaction. I will
return to this point later but first we will follow Ziegler
in the further development of his career.
In 1925-1926 he lectured at the University of Frankfurt
and in 1926 he joined Karl Freudenberg at Heidelberg.
Willstatter's prophecy that he wouldn't stay long in Heidelberg proved to be false: Ziegler showed a marked antipathy

to the political developments which were soon to appear and
in spite of the economic crisis and political climate he
spent ten happy years at Heidelberg. In this time the three
areas which were to attract his main interest took shape. It
was soon no longer possible to ignore Ziegler's scientific
achievements and in spite of his political opinions he was
offered the chair of chemistry at the University of Halle in
1936 where he became director of the Halle Institute. In
1943 he was offered the directorship of the Kaiser-Wilhelm-
Institut für Kohlenforschung in Mülheim a.d. Ruhr. Ziegler's
initial reaction was negative since he could see no connection between his interests and that of coal research.
However, the board of directors showed considerable foresight
and allowed him to choose his own research area. His interpretation of coal research as being the investigation of the
carbon-compounds in its widest sense did not always meet with
the full approval of the financing parties in the Ruhrgebiet
but the critical voices were silenced after the developments
of 1952.

I have mentioned that Ziegler was engaged mainly in
three research fields: two of these, "Zur Kenntnis des
'dreiwertigen' Kohlenstoffs" (concerning three-valent carbon)
and "Untersuchungen über alkaliorganische Verbindungen (concerning the organoalkalimetal compounds), have their origin
in his "Habilitations" thesis while the third, "Über viel-
gliedrige Ringe" (concerning many membered rings) was
developed in the thirties. The second theme was later
generalized to "Metallorganische Verbindungen" (concerning
organometallic compounds). In the following I shall briefly
discuss each of these topics.

THREE-VALENT CARBON

In 1923 Ziegler wrote: "The following is intended to be
the first of a number of publications in which an attempt is
made to identify the factors responsible for the dissociation
of substituted ethane derivatives". Ziegler's concern with
this problem during the years between 1923 and 1950 resulted
in many publications. By systematic alteration of the substituents in hexa-substituted ethane derivatives e.g., the
replacement of an unsaturated group by a saturated one,
informations were obtained on the role of steric and electronic factors in the dissociation process. A method
developed by Ziegler and Schnell - the so-called ether-method
- enabled the substituents to be widely varied.

2

The presence of aliphatic substituents hindered the determination of the dissociation equilibrium on account of the
instability of the radical dissociation products which
undergo further disproportionation.

In order to obtain information on the bond strength it
was, therefore, necessary to proceed indirectly and the rate
of decomposition of the ethane derivatives was first determined. For this purpose trapping reactions involving iodine,
NO, oxygen, and quinone were found to be particularly suitable. From the temperature dependence of the rates of decomposition it was possible to determine the heats of dissociation and activation energies. In this way the activation
energies of the following examples were determined.

3

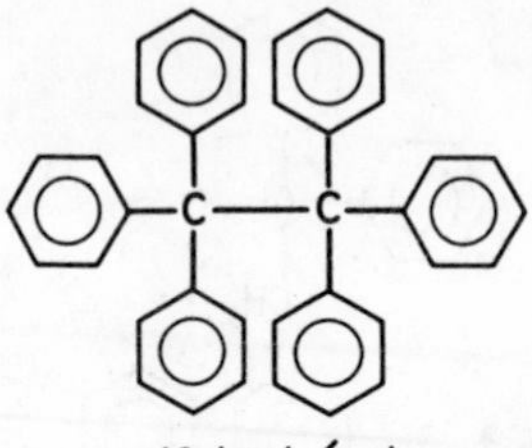

19 kcal/mol

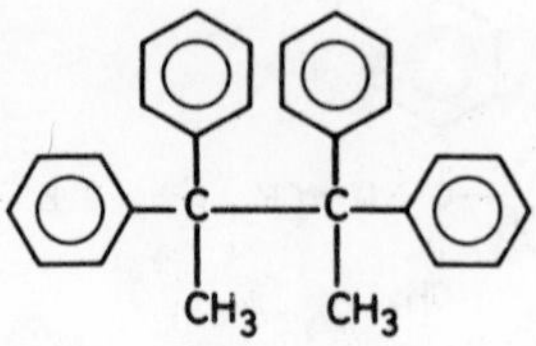

30 kcal/mol

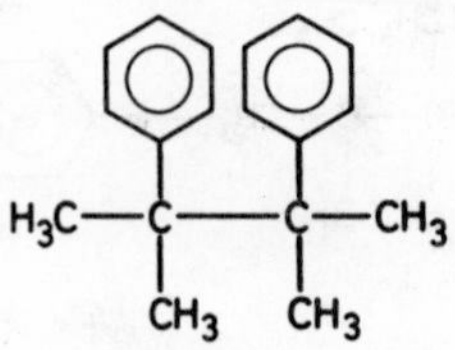

50 kcal/mol

Good agreement was obtained between these experimentally determined values and those obtained by E. Hückel at about the same time using a theoretical approach.

A colleague from Princeton once remarked that Ziegler was one of the very first "physical organic chemists". His work with radicals shows that at an early date he introduced the systematic use of physical-chemical methods into organic chemistry. In recent years this area has attracted renewed interest and on the basis of UV- and NMR-spectral evidence the dimerization of the trityl radical should be reformulated.

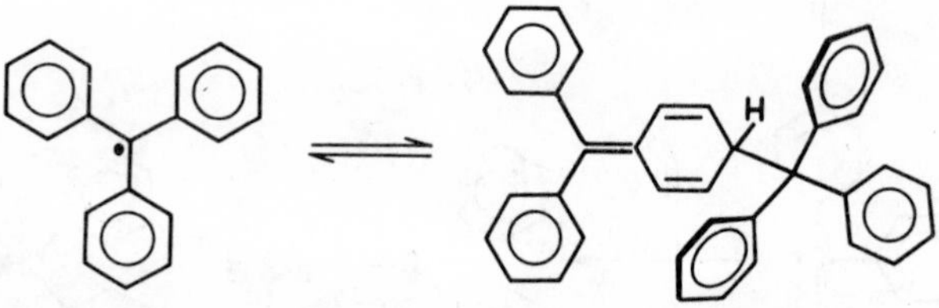

At the outset of the investigations on three-valent carbon, Ziegler was able to isolate two compounds having a very pronounced radical character: tetraphenylallyl and pentaphenylcyclopentadienyl. Molecular weight measurements show that these are dissociated to 80% and 100% respectively, which values are several powers of ten higher than that of Gomberg's triphenylmethyl (dissociation constant 2.2×10^{-4} at 24°C).

$$C_6H_5-C=CH-\overset{\bullet}{C}\overset{C_6H_5}{\underset{C_6H_5}{|}}$$

(Structure: diphenyl substituted at left, $C=CH$, radical carbon with two C_6H_5 groups)

(Right structure: cyclopentadienyl radical with five C_6H_5 groups)

ORGANO-ALKALI METAL COMPOUNDS

From the start Ziegler's work with radicals was closely associated with his investigations of the organo-alkali metal compounds and in this there is a certain analogy with the work of his friend, G. Wittig, who was also a student of Karl von Auwers. The alkali metals can act both as a radical trap or, if treated with the appropriate chlorate, as a starting point for the preparation of radicals.

$$R\bullet \; + \; Na \longrightarrow RNa$$

$$RNa \; + \; RClO_4 \longrightarrow 2\,R\bullet \; + \; NaClO_4$$

The organometallic compounds were prepared by an ether-cleavage reaction which was found to occur particularly smoothly if the ethereal carbon atoms were substituted by aromatic groups. In this way it was possible to prepare phenyl-*n*-propyl potassium.

$$C_6H_5-\overset{CH_3}{\underset{CH_3}{\overset{|}{\underset{|}{C}}}}-OCH_3 \; + \; 2\,K \longrightarrow C_6H_5-\overset{CH_3}{\underset{CH_3}{\overset{|}{\underset{|}{C}}}}-K \; + \; KOCH_3$$

This compound is of some significance since it was shown to add to the C=C double bond of stilbene.

$$C_6H_5-\overset{CH_3}{\underset{CH_3}{\overset{|}{\underset{|}{C}}}}-K \; + \; \overset{HC-C_6H_5}{\underset{HC-C_6H_5}{||}} \longrightarrow C_6H_5-\overset{CH_3}{\underset{CH_3}{\overset{|}{\underset{|}{C}}}}-\overset{C_6H_5}{\underset{H}{\overset{|}{\underset{|}{C}}}}-\overset{C_6H_5}{\underset{H}{\overset{|}{\underset{|}{C}}}}-K$$

This particular reaction was described by Ziegler in his 1925 "Habilitations" thesis and is the prototype of the organo-metallic syntheses which were later to become of such importance in organoaluminum chemistry. The reaction was also the key to an understanding of the mechanism of the polymerization of butadiene with sodium to give the so-called "Buna rubber." An initial indication of the mechanism of this polymerization reaction had been obtained by reacting butadiene with butyllithium (which was prepared by treating butyl chloride with lithium, a general reaction discovered by Ziegler and Colonius in 1930).

In ether, butyl lithium rapidly polymerizes butadiene. However, if the reaction is terminated at the outset by adding water, then a mixture of unsaturated hydrocarbons having the general formula $C_4H_9-(C_4H_6)_n-H$ (n = 1 to 6) is obtained. The first step in this organometallic synthesis was illustrated by reacting triphenylmethyl sodium with butadiene in the presence of diethylamine.

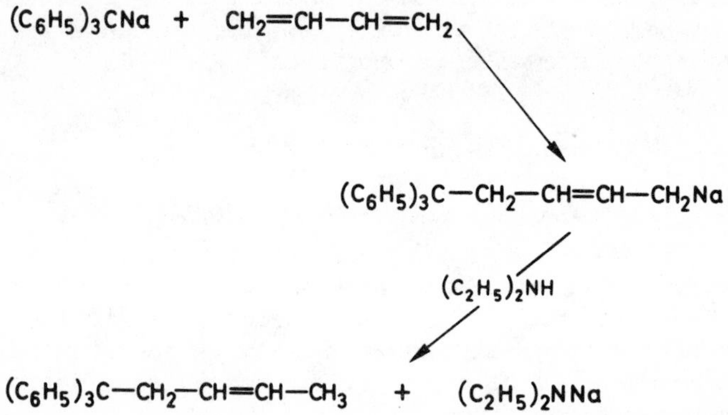

$(C_6H_5)_3CNa$ + $CH_2=CH-CH=CH_2$

$(C_6H_5)_3C-CH_2-CH=CH-CH_2Na$

$(C_2H_5)_2NH$

$(C_6H_5)_3C-CH_2-CH=CH-CH_3$ + $(C_2H_5)_2NNa$

An analogous proof of the mechanism of the sodium induced polymerization of butadiene was also ultimately obtained.

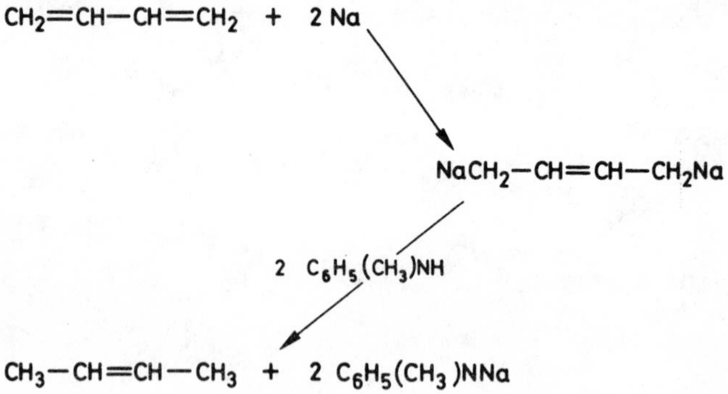

$CH_2=CH-CH=CH_2$ + 2 Na

$NaCH_2-CH=CH-CH_2Na$

2 $C_6H_5(CH_3)NH$

$CH_3-CH=CH-CH_3$ + 2 $C_6H_5(CH_3)NNa$

MANY-MEMBERED RINGS

Before turning my attention to what has become known as
Ziegler-chemistry *per-se*, I would like to mention his work on
large ring systems as well as the Wohl-Ziegler bromination.
Interest in ring systems larger than the six-membered
ring systems had been stimulated by Ruzicka's work in the
years between 1926 and 1933 on the naturally occurring
perfumes such as musk. Ruzicka had been able to prepare 6-
to 18-membered cyclic ketones by thermal degradation of the
thorium salts of the appropriate α,ω-dicarboxylic acids and
had observed that the yields reached a minimum around the
ten-membered rings. In 1933 appeared the first publication
by Ziegler, Eberle and Ohlinger describing in detail the synthesis of large ring systems. The basis of these investigations became known as the Ruggli-Ziegler dilution principle
and involved a method devised by Ruggli some twenty years
earlier for preparing cyclic amides: high dilution was found
to suppress intermolecular reactions in favor of intramolecular ring formation. This work had apparently been ignored
until Ziegler developed a general method which enabled 14-
to 33-membered rings to be synthesized in 60% - 80% yield -
a vast improvement on the 10% yields obtained by Ruzicka.
The yield of the 9- to 13-membered rings was considerably
less; in the case of the 12- and 13-membered rings, 10-15%
and in the case of the 9-, 10- and 11-membered rings only
traces were obtained. The basis of Ziegler's synthesis was
the cyclization of α,ω-dinitriles using lithium diethylamide,
lithium ethylanilide or sodium methylanilide.

These results demonstrated once again that ten-membered
rings are particularly difficult to synthesize. In addition
a second minimum was suggested to occur for the 25- and 26-
membered rings while a periodicity was observed in the yields
of the odd- and even-membered rings.

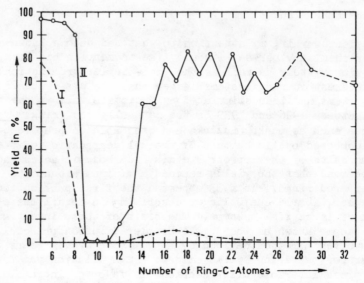

Number of Ring-C-Atomes ⟶

A new interest in ring systems developed around 1947 when
Ziegler discovered that at above 200° butadiene can be converted in 15% yield into 1,5-cyclooctadiene. Starting from
this easily prepared eight-membered ring it was possible,
using classical methods, to prepare for the first time
strained eight-membered cyclic olefins having a *trans*-double
bond.

Yet a further reaction which is associated with Ziegler
is the bromination of allylic olefins using N-bromosuccini-
mide. This reaction became known as the Wohl-Ziegler bromination and has become a standard reaction in organic synthesis.

ORGANOMETALLIC COMPOUNDS

The most significant part of Ziegler's success began
with the simple observation that lithium alkyls decompose at
elevated temperatures; for example, during distillation.

$$Li-C_2H_5 \rightleftharpoons LiH + CH_2{=}CH_2$$

8

The question arose as to whether it would be possible to reverse this reaction by applying pressure and thereby introduce a new method for synthesizing lithium alkyls. But all these experiments were unsuccessful. In 1949, Ziegler and Gellert reacted LiAlH$_4$ - discovered in 1947 by Finholt, Bond and Schlesinger - dissolved in ether with ethylene under pressure at 140° and obtained higher α-olefins as the result of a stepwise organometallic reaction. It soon became apparent that aluminum alkyls were even more reactive and from the reaction of triethyl aluminum with ethylene under pressure at 100° it was possible to isolate higher aluminum alkyls having an alkyl group which, in extreme cases, resulted from the reaction of 100 ethylene molecules.

$$Al \underset{C_2H_5}{\overset{C_2H_5}{\underset{|}{-}C_2H_5}} + x\ C_2H_4 \longrightarrow Al \underset{(C_2H_4)_c-C_2H_5}{\overset{(C_2H_4)_a-C_2H_5}{\underset{|}{-}(C_2H_4)_b-C_2H_5}}$$

$$x = a + b + c$$

This reaction became known as the "Aufbaureaktion" (insertion reaction) and its extent is limited by the so-called "Verdrängungsreaktion" (elimination reaction) which produces α-olefins.

$$al-CH_2-CH_2-R \longrightarrow alH + CH_2=CH-R$$

$$al = \tfrac{1}{3}Al$$

This "Aufbaureaktion" forms the basis of a method for preparing primary alcohols having a 12- to 16-carbon atom chain and which are of importance in the preparation of biodegradable detergents. The production figures run at the moment at 120,000 - 150,000 tons per year.

$$al-C_2H_5 \ + \ x \, C_2H_4 \longrightarrow al-(C_2H_4)_x-C_2H_5 \xrightarrow{\frac{1}{2} O_2}$$

$$al-O-(C_2H_4)_x-C_2H_5$$

$$\xrightarrow[-alOH]{H_2O}$$

$$HO-(C_2H_4)_x-C_2H_5$$

$$x \, = \, \approx 6-8$$

The dimerization of propylene to 2-methylpent-1-ene using aluminum alkyls involves both the insertion and elimination reaction and is the initial step in the preparation of isoprene by the Goodyear Scientific Design Process which is used to synthesize approximately 100,000 tons of isoprene per year.

$$al-C_3H_7 \ + \ CH_2\!\!=\!\!CH-CH_3$$

$$al-CH_2-\underset{\underset{CH_3}{|}}{CH}-C_3H_7$$

$$al-C_3H_7 \ + \ CH_2\!\!=\!\!\underset{\underset{CH_3}{|}}{C}-C_3H_7 \xleftarrow{\ C_3H_6\ }$$

These processes are all based upon the discovery of a simple synthesis of aluminum trialkyls which is related to the initial work with the lithium alkyls.

$$Al \ + \ \frac{3}{2}H_2 \ + \ 2\,C_2H_4 \longrightarrow (C_2H_5)_2\,AlH \xrightarrow{\ C_2H_4\ } Al(C_2H_5)_3$$

It is now common-place to produce various aluminum alkyls in quantities of 1000's of tons per year. However, this development would certainly not have taken place so rapidly had not Ziegler demonstrated at an early stage that it is even possible in a research institute to prepare in a continuous process large quantities of these spontaneously inflammable materials which explode with water. In this particular case it was necessary to prepare, not the "Ziegler-Kilogramm" (which was a source of amusement and some discomfort to his coworkers), but a "Ziegler-ton".

The "Aufbaureaktion" involving ethylene had received much attention and had always led to more or less the same result – the formation of a higher aluminum trialkyl. Completely unexpectedly one experiment in 1953 produced exclusively butene-1. After a search of several weeks, Ziegler and Holzkamp discovered that traces of nickel which had been produced during the cleaning of the autoclave cause the "Aufbaureaktion" to terminate after one insertion step and the result is the elimination of butene-1, the dimer of ethylene.

$$al{-}C_2H_5 \;+\; C_2H_4$$

$$al{-}CH_2{-}CH_2{-}C_2H_5$$

$$al{-}C_2H_5 \;+\; CH_2{=}CH{-}C_2H_5 \;\xleftarrow[C_2H_4]{Ni}$$

A systematic investigation of this effect led in the middle of 1953 to the discovery of the Ziegler catalysts which were shown to convert ethylene and other α-olefins into high molecular weight materials. Up until this time ethylene was considered to be extremely difficult to polymerize – pressures of 1000 – 2000 atm. and temperatures of 200° were necessary – and hence a process capable of polymerizing ethylene at atmospheric pressure to produce a product in many ways superior was a true sensation of tremendous technical consequence. The term Ziegler catalyst was originated by G. Natta, who as a consultant for the firm Montecatini, had already in 1953, access to all the information on the new catalysts as a result of a licence which Ziegler had granted to Montecatini in 1952. Natta subsequently discovered that the catalysts function stereospecifically and lead in particular to isotactic polymers.

In general Ziegler catalysts are formed by the reaction of a transition metal compound with an organometallic compound. For example, $TiCl_4$ reacts with $(C_2H_5)_2AlCl$ to form a fine suspension of $TiCl_3$ which, in association with the organoaluminum component, produces a highly active catalyst. The most active Ziegler catalysts are based nowadays upon compounds of titanium, vanadium and cobalt in combination with organoaluminum compounds. They are used to prepare polyethylene, polypropylene, poly-4-methyl-pent-1-ene, ethylene-propylene rubber, *cis*-1,4-polybutadiene, and *cis*-1,4-polyisoprene.

$$TiCl_4 + (C_2H_5)_2AlCl \longrightarrow$$

$$[TiCl_3 C_2H_5] + AlCl_2C_2H_5$$

$$\downarrow -\cdot C_2H_5$$

$$TiCl_3$$

It is estimated that today the value of the materials produced by processes based on Ziegler's discoveries throughout the world is 3 Mrd DM per year. The industrial evaluation of the original discovery by K. Ziegler, E. Holzkamp, H. Breil, H.G. Gellert and H. Martin has of course only been possible as the result of the efforts of innumerable chemists and engineers throughout the world and it is hardly possible to estimate the number of publications and patents which are concerned with the preparation and application of the Ziegler catalysts. Even so, the mechanism of the reaction is still not understood in all its details.

The description which I have given of the scientific achievements of Karl Ziegler are hardly compatible with the picture of a scientist living in an ivory tower. Initially organometallic chemistry was a curiosity and it was only in Ziegler's hands that it developed into an area of such significance. Karl Ziegler combined the abilities of a brilliant chemist with a talent for recognizing the technical possibilities of the results of fundamental research. He also possessed the energy and acumen to foster the commercial exploitation of his own discoveries. As a result of this fortunate combination, Karl Ziegler was able on the 1st of October 1968, his 25th anniversary at the Institute, to create the "Ziegler-Fund" which he endowed with a starting capital of 40 million DM.

In spite of this, Ziegler was basically an innovator. He always had clear objectives and saw in the success of an experiment not mainly a confirmation of his own ideas, but rather the possibility of obtaining new knowledge. His guiding principle was that it is not possible to anticipate something which is really new; this can only be discovered by experiment.

Ziegler's successes led to many honours and awards. Among others he was awarded honorary doctorates at the

Universities of Heidelberg and Gießen as well as at the
Technical Universities of Hannover and Darmstadt; he was
President of the German Chemical Society during the difficult
years from 1946 to 1951 and was elected an honorary member;
in 1968 he was elected an honorary senator of the Max-Planck-
Society. The culmination of these honours was the award of
the Nobel Prize for Chemistry in 1963 which he shared with
G. Natta. This was certainly a fitting award, made in the
spirit of Dr. Nobel's, with which to honor those whose achievements have been of greatest benefit to mankind.

The many students and scientists from all over the world
who worked with Ziegler at various times were able to learn
under his guidance to tackle a problem with enthusiasm and to
seek the answer by systematic investigation. In addition it
was one of Ziegler's principles to keep an eye open for
unexpected developments and not to neglect new phenomena as
irrelevant to the main project. However, not only was it
possible to learn from his scientific approach, but also from
his philosophy to life whereby he had a talent for finding
simple solutions to complicated problems.

Outside of the laboratory, Ziegler found relaxation in
music and in collecting paintings. He was also a keen
climber and I remember as a young assistant having had the
opportunity of accompanying him to the Alps where I discovered
to my amazement that for the length of a summer vacation he
hardly mentioned chemistry. This ability also undoubtedly
contributed to his success. He has himself described the key
to his success: "I have had the good fortune, through an
early observation, to stumble upon a new and fertile area in
my youth and as a result I needed to do nothing more than to
satisfy my scientific curiosity. Thus it can be understood
when I have said that my aim has been exclusively to do that
which gives me pleasure. I have never tried to exhaust the
possibilities of my newly won knowledge. I was satisfied to
open the door".

Although he no longer lives, the life and work of Karl
Ziegler can act as an example to us all.

A Few Considerations on Stereoregular Propylene Polymerization

ADOLFO ZAMBELLI

*Istituto di Chimica delle Macromolecole del CNR,
Via Alfonso Corti, 12 - 20133 Milano, Italy*

1) INTRODUCTION

As pointed out in a previous paper(1), the structure of
a macromolecule is a record   of what happened during the
polymerization reaction. Hence, by combining the use of
spectroscopic analysis techniques on polymers, mainly n.m.r.,
i.r., and isotopic substitution, it has been possible to gain a rather detailed understanding of the stereochemistry of the polymerization of α-olefins in the presence
of Ziegler catalysts. Therefore, the main aim of this paper
is to try to construct, from the stereochemical data obtained
during the past few years, a possible stereoregulation mechanism, valid for both isotactic and syndiotactic specific polymerizations of propylene(2).
    Taking into account that it has been well established
that polymerization occurs by the insertion of the monomer on
reactive transition metal (M) carbon bonds(3-5)

$$M-P + C_3H_6 \longrightarrow M-(C_3H_6)-P$$

(where P is the growing chain bonded to M), the stereochemistry is concerned with deciding:
(1) what is the orientation of the monomer in the insertion
     step

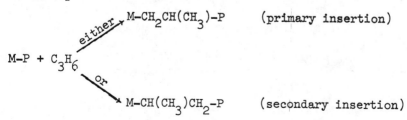

M-P + C_3H_6 —— either → M-CH$_2$CH(CH$_3$)-P   (primary insertion)

—— or → M-CH(CH$_3$)CH$_2$-P   (secondary insertion)

(2) what is the stereochemical mechanism of addition to the
double bond (*cis* addition or *trans* addition)
(3) what is the role of the possible chirality of M and of
the chirality of the last unit of the growing chain end
in steric control.

2) *ISOTACTIC POLYMERIZATION*

The data concerning both the terminal unsaturations of
dead macromolecules(6) and the nearly stoichiometric reaction
between monomer and catalyst(7) seem to show that isotactic
polymerization proceeds by the primary insertion,

$$M-P + C_3H_6 \longrightarrow M-CH_2CH(CH_3)-P$$

Natta, Farina et al.(8) showed that polymerization of
cis-($1d_1$)-propylene and of *trans*-($1d_1$)-propylene allows one
to obtain two different stereoregular di-isotactic polymers;
Miyazawa(9) established that the polymer of *cis*-($1d_1$)-
propylene is *erythro*-di-isotactic, while the polymer of
trans-($1d_1$)-propylene is *threo*-di-isotactic (Fig. 1).

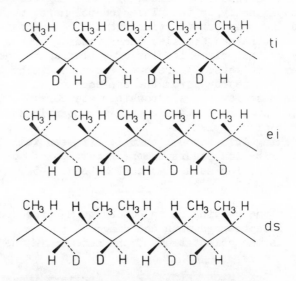

*Fig. 1. Three stereoregular polymers
have been obtained from cis and trans
$1d_1$ propylene: threo-diisotactic (ti),
erythro-diisotactic (ei), and disyndio-
tactic (ds).*

So, it was possible to conclude that the mechanism of addition to the double bond is *cis*.

The isotactic steric control seems to be due to the chirality of the metal atom of the catalyst centers because, with typical isotactic specific catalysts, the reaction is stereospecific   when the last unit of the growing chain end is achiral. In fact, in ethylene-propylene copolymers, prepared with the above catalysts, all isolated ethylene units (those having two propylene first neighbour units) lie in the same steric environment (reasonably meso)(1) (Fig. 2.).

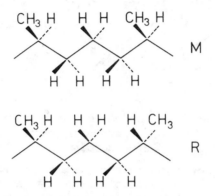

Fig. 2. Isolated ethylene unit in a vynil chain could be found in meso (M) or racemic (R) environment.

It is evident that ethylene both in meso and racemic environment should be present, if the stereospecific addition of the n-th unit is   dependent upon the presence of a chiral carbon atom in the n-1th unit.

3)   SYNDIOTACTIC POLYMERIZATION

Syndiotactic polymerization, also, occurs by *cis* addition to the double bond(10). Due to the degeneration of the structures of *erythro* and *threo* di-syndiotactic polymers of 1,2 disubstituted monomers, this result was achieved only by resorting to the comparison of the structures of copolymers of perdeuteropropylene with *cis*- and *trans*-(1d₁)-propylene. In fact, the *erythro* units are *perse* distinguishable from the *threo* units, but when they are linked one to another in a syndiotactic way, *erythro* di-syndiotactic and *threo*-disyndiotactic chains having the same structure are obtained(8) (Fig. 3).

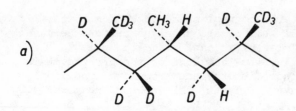

a) *threo* (or *gauche*) di-syndiotactic copolymer

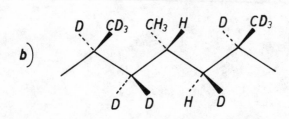

b) *erythro* (or *trans*) di-syndiotactic copolymer

Fig. 3. No degeneration of the erythro and threo units takes place in perdeuteropropylene - $1d_1$ propylene syndiotactic copolymers.

In ethylene-propylene copolymers, prepared with syndiotactic specific catalysts both meso and racemic isolated ethylene units were formed (Fig. 2); this fact shows that, in this case, the steric control is due to the chirality of the substituted carbon of the last unit of the growing chain end(11).

The structure of ethylene-propylene copolymers revealed also that the propylene insertion, in syndiotactic homopolymerization, is essentially secondary(12) because of steric reasons(13). In fact, when the last unit of the growing chain end is a propylene unit arranged as shown below,

$$M-CH(CH_3)CH_2-P+C_3H_6 \longrightarrow M-[CH(CH_3)CH_2]_2-P$$

propylene insertion is mostly secondary while it is mostly primary when the last unit is an ethylene(13).

$$M-CH_2CH_2-P+C_3H_6 \longrightarrow M-CH_2CH(CH_3)CH_2CH_2-P$$

The driving force for the syndiotactic pathway of the polymerization is in the range of 1.5 - 2.5 kcal/mole (14).

4) *STERIC AND ARRANGEMENT DEFECTS*

Many boiling *n*-heptane insoluble fractions of polypropylene have an almost pure isotactic structure. No defects in the chemical arrangement are detectable in such polymers

except 1-2% isolated syndiotactic diads(15-17)

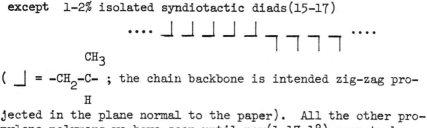

CH₃

($\rfloor$ = -CH$_2$-C- ; the chain backbone is intended zig-zag pro-
H

jected in the plane normal to the paper). All the other propylene polymers we have seen until now(1,17,18) seem to be
block-like in the sense that homo diads sequence is preferred over hetero diads sequence.* For instance, the
prevailingly syndiotactic polymers obtained with homogeneous
vanadium-based catalysts seem to consist of rather regular
syndiotactic stereoblocks and of stereoirregular blocks(17).

Such incompletely stereoregular polymers also contain
some defects in the chemical arrangement, i.e. head-to-head
($\lfloor$__$\rfloor$) and tail-to-tail ($\rfloor$ $\lfloor$) arranged units(13). Roughly
speaking, the lower is the stereoregularity, the greater is
the content of head-to-head and tail-to-tail arranged units
of a given polymer.

5) SEQUENCE DISTRIBUTION IN ETHYLENE-PROPYLENE COPOLYMERS

This topic could seem a digression from the subject, but
steric control in propylene homopolymerization, and the control of distribution of the different units in ethylene-
propylene copolymerization seem to be strictly correlated(20).
In fact, syndiotactic specific catalytic complexes give
copolymers with an alternating-like sequence distribution
($r_1r_2 < 1$), while aspecific or isospecific complexes give a
almost Bernoullian sequence distribution ($r_1r_2 \backsimeq 1$).

6) A FEW CONSIDERATIONS ON THE STERIC CONTROL

Syndiotactic specific steric control is easily understood if *cis* addition and secondary monomer insertion are
considered. In fact these suggest a four-member active
complex as shown in Fig. 4a. This would ensure syndiotactic
propagation because of its greater stability in comparison
with the isomer complex having the methyl groups (of the

(*) The most commonly used description of steric structure
is that proposed by Frish, Mallows and Bovey based on m
(isotactic) diads and r (syndiotactic) diads(19).

monomer and of the last unit) in *cis* position (7,21)(Fig. 4b).

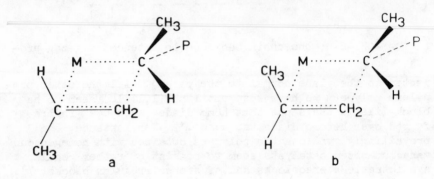

Fig. 4.   *The active state a (trans) should be more stable*
than b (cis).

It must be noted that the approach of the monomer to the
reactive metal-carbon bond ought to occur from the less
hindered side, and that the greater stability of the *trans*
complex should arise because the substituted carbon atom of
the last unit is involved in the formation of the four member
ring, so that rotation is not allowed.   The regular arrangement of the unit should arise from the comparatively high
energy of the complexes substituted in vicinal positions
(Fig. 5).

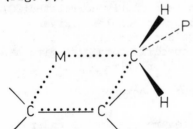

Fig. 5.   *Vicinal substitution*
in the active state should
oppose tail to tail
arrangement.

In any event, after an arrangement error, which dis-
places the methyl group of the last unit farther away, both
the driving forces of the syndiospecific control and of the
secondary insertion should be lost.   The same would obviously
occur in the copolymerization after the insertion of an
ethylene unit (Fig. 6).   In fact, primary insertion seems to
be preferred after such a methyl displacement, probably due to
the hindrance of the substituents of M (13).   This fact
would explain the blocky nature of prevailingly syndiotactic
homopolymers, which should consist of syndiotactic and

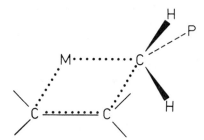

Fig. 6. When chain ends in a methylene groups, the methyl group of the approaching monomer occupies any of the four positions shown with comparable frequencies.

stereoirregular blocks connected through head-to-head and tail-to-tail arranged units.

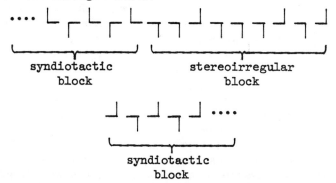

In other words, secondary insertion on the substituted carbon-ending chain would essentially involve syndiotactic propagation. Primary insertion of one monomer molecule on a substituted carbon-ending unit would generate one CH_2-ending chain and lessen the steric control driving force; primary insertion of monomers on CH_2-ending chains would not be syndiotactic specific, and somewhat preferred over secondary insertion. Secondary insertion of one monomer molecule on a CH_2-ending chain would restore the syndiotactic control and the secondary insertion driving force.

As to the ethylene-propylene copolymerization, it is easy to foresee that the ratio between pre-exponential factors for the addition of ethylene and propylene to one substituted carbon-ending chain would be near 4/1, and near 1/1 for the addition to one CH_2-ending unit (Fig. 6). This, perhaps, is the reason why, with syndiotactic specific catalysts, an alternating-like sequence distribution of the different units is found.

Isotactic propagating would arise: (1) if the metal atom of the catalyst complexes bears such bulky substituents as to forbid secondary insertion; (2) the bulkiness of such

substituents is different, so that constant presentation of
the monomer is forced; (3) the monomer always approaches the
reactive metal-carbon bond from the same side. Such requirements are met by an activated complex as in Fig. 7, where X
and Y are two substituents of M (Y bulkier than X) lying
above and below the plane of the four-membered ring of the
active complex.

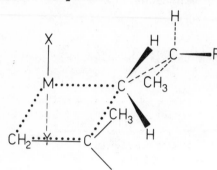

*Fig. 7. Steric hindrance
above and below M should be
determining in order that
isotactic propagation may
take place.*

Monomer insertion instead of *cis* ligand migration is
supposed to occur. This fact would arise from the actual
geometry of the complexes and substantially it coincides with
the *cis* ligand migration, followed by rearrangement, proposed
by Cossee and Arlman(22) for the heterogeneous catalysts.

It is worth noting that such a mechanism allows one to
foresee the same ratio, 4/1, between the pre-exponential
factors in the ethylene and propylene addition to growing
chains ending either in a propylene or in an ethylene unit:
this gives a possible explanation of the value of the product
of the reactivity ratios in the ethylene-propylene copolymerization with such isotactic specific catalysts ($r_1 r_2 \simeq 1$).

Clearly, when changing the steric environment above and
below M, no clear cut prediction can be made about isotactic specific and syndiotactic specific catalyst centers. In
principle, both isotactic specific (or not stereospecific)
and syndiotactic specific propagation could take place on
some catalyst complexes, depending on which monomer insertion
(primary or secondary)   actually occurs. This could be
an explanation for the stereoblock polymers that are obtained
in some instances.

Occasional inversions of the configuration of the isotactic specific catalyst centers during the growth of one
macromolecule, could account(21) for the isolated r diads
in the highly isotactic polymers.

7) CONCLUSION

It seems to us that many of the main stereochemical features of the stereospecific polymerization of α-olefins can be fitted fairly satisfactorily into a quite classical mechanistic picture, involving a concerted *cis* addition to the double bond. The type of insertion of the monomer, secondary instead of primary, should be the key to the occurrence of the syndiotactic specific propagation instead of the isotactic specific or not stereospecific.

The steric environment of the metal of the catalyst centers is the determining factor for the isotactic (instead of non stereospecific) propagation to occur through primary insertion.

LITERATURE CITED

(1) A. Zambelli, "NMR Basic Principles and Progress", Vol. 4, 101 Berlin-Heidelberg-New York: Springer Verlag 1971.
(2) J. Boor, Jr., *J. Polymer Sci. D2*, 115 (1967).
(3) C.F. Feldman, and E. Perry: *J. Polymer Sci. 46*, 217 (1960).
(4) G. Natta, and I. Pasquon. *Adv. Catal., Vol. 11*, Ch. 11, p. 1; New York: Academic Press, 1959.
(5) A. Zambelli, G. Natta, and I. Pasquon, *J. Polymer Sci., C4*, 411 (1963).
(6) P. Longi, G. Mazzanti, A. Roggero, and M.P. Lachi, *Makromol. Chem., 61*, 63 (1963).
(7) Y. Takegami and T. Okazaki, *Bull. Chem. Soc., Japan 42*, 1060 (1969).
(8) G. Natta, M. Farina, and M. Peraldo, *Chim. Ind. (Milano) 42*, 255 (1960).
(9) T. Miyazawa, and T. Ideguchi, *J. Polymer Sci., B1*, 389 (1963).
(10) A. Zambelli, M.G. Giongo, and G. Natta, *Makromol. Chem. 112*, 183 (1968).
(11) A. Zambelli, G. Gatti, C. Sacchi, W.O. Crain, Jr., and J.D. Roberts, *Macromolecules 4*, 475 (1971).
(12) Y. Takegami, and T. Suzuki, *Bull. Chem. Soc. Japan, 42*, 848 (1969).
(13) A. Zambelli, C. Tosi, and C. Sacchi, *Macromolecules, 5*, 649 (1972).
(14) Unpublished data from our laboratory.
(15) A. Zambelli, A.L. Segre, M. Farina, and G. Natta, *Makromol. Chem., 110*, 1 (1967).

(16) F. Heatley, R. Solovey, and F.A. Bovey, *Macromolecules*, *2*, 619 (1969).
(17) A. Zambelli, D.E. Dorman, A.I. Richard, and F.A. Bovey, *Macromolecules*, *6*, 925 (1973).
(18) A. Zambelli, L. Zetta, C. Sacchi, and C. Wolfsgruber, *Macromolecules 5*, 440 (1972).
(19) H.L. Frish, C.L. Mallows, and F.A. Bovey, *J. Chem. Phys.* *45*, 1565 (1966).
(20) A. Zambelli, A. Léty, C. Tosi, and I. Pasquon, *Makromol. Chem.*, *115*, 73 (1968),
(21) A. Zambelli, and C. Tosi, *Adv. in Polymer Sci. Vol. 15*, in press.
(22) E.J. Arlman and P. Cossee, *J. Catalysis 3*, 99 (1964).

Stereoselection and Stereoelection in α-Olefin Polymerization

P. PINO, A. OSCHWALD

Eidgenossische Technische Hochschule
Zurich, Switzerland

F. CIARDELLI, C. CARLINI, E. CHIELLINI

Istituto di Chimica Organica Industriale
University of Pisa, Italy

1)   INTRODUCTION

The discovery of "stereoselective"[+] polymerization of
racemic α-olefins was originated from investigations of their
stereospecific polymerization(1-3). It was observed(4) that
by polymerization of racemic 4-methyl-1-hexene [(R)(S)4-MH]
with typical Ziegler-Natta catalyst, a highly crystalline high
melting polymer was obtained. This result was rather unexpected since polymerization of a racemic α-olefin should lead
to a random copolymer of the two antipodes (Scheme 1a). The
steric irregularity due to the random distribution of the (R)
and (S) asymmetric carbon atoms[++] in the lateral chain should
be sufficient to hinder crystallization, even when the principal chain is completely isotactic. In order to explain the
high crystallinity of the isotactic polymer of (R)(S)4-MH,
two hypotheses were formulated:

[+]   In the classical organic stereochemistry a process is called
stereoselective when it produces a diastereomer of a given
structure in considerable predominance over all the other
possible diastereomers(1). The above definition can be extended to an oligomerization or polymerization process giving as a
limit case isotactic macromolecules each of [(R) or (S) monomeric units.]
[++] With (R) or (S) asymmetric carbon atom we mean asymmetric
carbon atom having (R) or (S) absolute configuration.

SCHEME 1.

Polymerization of racemic 4-methyl1-1-hexene to: (a) random copolymer of the two antipodes; (b) mixture of homopolymers of the two antipodes.

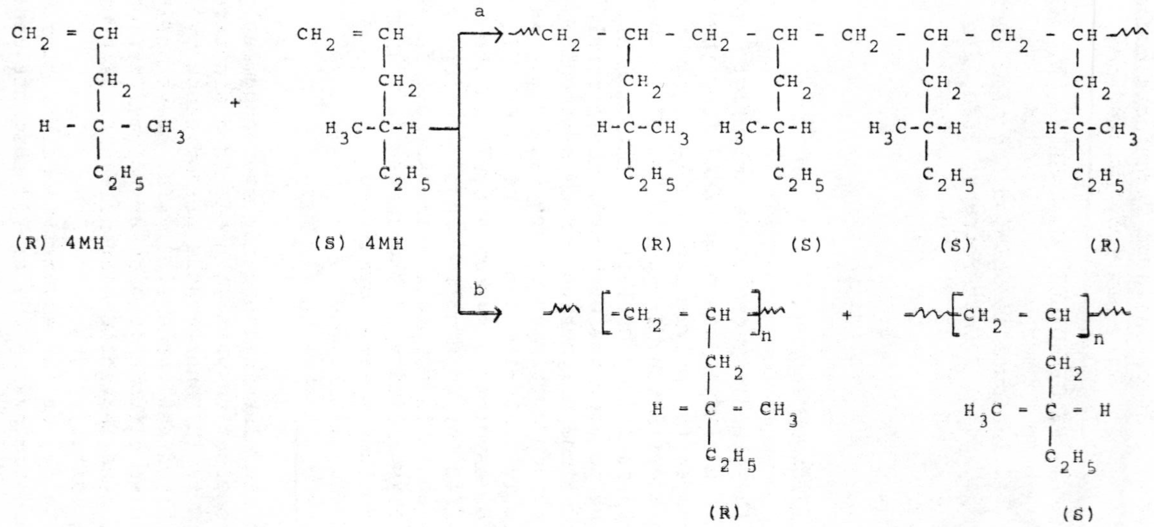

$CH_2 = CH$
CH_2
$H - C - CH_3$
C_2H_5

(R) 4MH

$+$

$CH_2 = CH$
CH_2
$H_3C - C - H$
C_2H_5

(S) 4MH

a →

$\sim\sim CH_2 - CH - CH_2 - CH - CH_2 - CH - CH_2 - CH - CH_2 - CH_2 - CH \sim\sim$
CH_2 CH_2 CH_2 CH_2 CH_2
$H-C-CH_3$ H_3C-C-H H_3C-C-H $H-C-C-H$ $H-C-CH_3$
C_2H_5 C_2H_5 C_2H_5 C_2H_5 C_2H_5
(R) (S) (S) (R)

b →

$\sim\sim \left[=CH_2 = CH \right]_n$
CH_2
$H = C - CH_3$
C_2H_5
(R)

$+$

$\sim\sim \left[CH_2 = CH \right]_n$
CH_2
$H_3C = C = H$
C_2H_5
(S)

26

(1) The steric irregularities associated with the presence of a random distribution of (S) or (R) asymmetric carbon atoms in the lateral chains of each macromolecule are not sufficient to hinder the packing of the chains in the crystalline lattice.

(2) The polymerization process gives rise to a mixture of macromolecules, each of which consists exclusively or prevailingly of (R) or (S) monomeric units (Scheme 1b).

In the former case a separation of the polymer in fractions having optical activity of opposite sign is theoretically possible, but, particularly in the case of high molecular weight polymers, should yield fractions with very low optical activity, even if the separation is complete(5). On the contrary, in the latter case, fractions having optical activity of the same order of magnitude of the polymer obtained by polymerization of a single monomer antipode should be obtained if an efficient separation method is used.

By chromatography on an optically active support it was indeed possible to separate into fractions having optical activity of opposite sign(2,6) polymers of racemic α-olefins having an asymmetric carbon atom in the α or β position with respect to the double bond. No separation was observed when the asymmetric carbon atom is in the γ position with respect to the double bond(6). As far as the crystallinity of the poly-α-olefins obtained from the racemic monomers is concerned, both hypotheses 1 and 2 play an appreciable role, their relative importance being dependent mainly on the monomer structure.

The existence, at least in some cases, of a remarkable stereoselectivity in the racemic α-olefins' polymerization suggested a new way to investigate the stereochemistry of the polymerization processes. Stereoselectivity was observed also in the polymerization of alkylene oxides(7), Leuch's anhydride (7), vinyl ethers(8) and diolefins(9), but these topics will not be discussed in this review. Stereoselectivity requires peculiar polymerization systems capable of distinguishing between the antipodes of a racemic monomer. It is a unique example of copolymerization where two monomers, having identical reactivity, give rise to a mixture of homopolymers only for steric reasons. In other words the values of the reactivity ratios $r_S = k_{SS}/k_{SR}$ and $r_R = k_{RR}/k_{RS}$ are identical but markedly greater than one, where k_{SS} (or k_{SR}) indicates the rate constant for insertion of a S (or R) unit in a chain, the last monomeric unit of which is S, and k_{RR} (or k_{RS}) have the same meaning for a chain where the last inserted monomeric

27

unit is R. A large value of r_S and r_R can in principle be due to a very strong steric interaction between the last monomeric unit in the growing chain and the new monomer entering in the chain or/and to the existence of chiral catalytic centers regulating the insertion of a new monomer molecule between the growing chain and the catalyst (Scheme 2). When the first factor, as for instance in the case of Ziegler-Natta polymerization of α-olefins, can be disregarded, the existence of stereoselectivity can be assumed as a reasonable proof of the existence of chiral catalytic centers(10).

In the present paper, after a section devoted to the qualitative demonstration and quantitative evaluation of stereoselectivity in racemic α-olefins polymerization, and to the influence exerted by monomer structure and catalyst nature, some interesting applications will be reported including stereoelective[+] polymerization of one enantiomer and stereo-selective copolymerization. The possible origin of stereo-selectivity will be discussed to explain those factors res-ponsible for the peculiar stereoregulating properties of Ziegler-Natta catalysts.

2) *STEREOSELECTIVE POLYMERIZATION*

When a racemic monomer is polymerized to give a high molecular weight polymer, there is a prevalence of (R) or (S) monomeric units in most of the macromolecules(5). These should give positive or negative contribution to optical activity, the net result being an optically inactive polymer. If the distribution of (R) and (S) monomer units is random in all the macromolecules then the excess of (R) or (S) monomeric units in a polymer portion can be calculated(5). For instance for the case of poly-(R)(S)4-MH for monodisperse (R)-(S) co-polymer samples having DP 10, 100, 1000, or 2000, an average excess of (R) or (S) units of 26, 7, 2.5, or 2%, respectively, can be foreseen at most for 50% of the macromolecules.

[+] When a racemic compound undergoes a catalytic synthesis in the presence of an optically active catalyst and the conver-sion is not complete, kinetic resolution(11) of the starting compound takes place and in general a new optically active pro-duct is formed. The formation of this last is not well defined by the expression "kinetic resolution"; we have indicated this process with the term stereoelective polymerization as one anti-pode is preferentially chosen (elected). Actually the first example of preferential polymerization of one antipode from a racemic monomer has been reported by Tsuruta, Inoue and Furu-kawa(12) who called this process "asymmetric-elective poly-merization".

SCHEME 2.

Steps of the anionic coordinated polymerization of a chiral α-olefin.

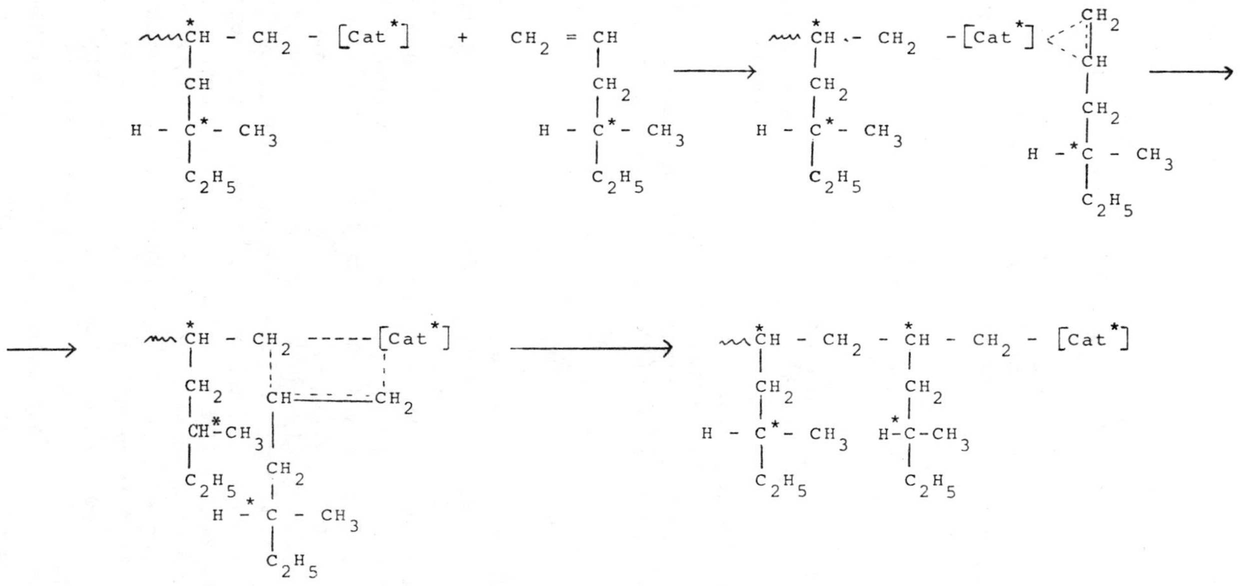

$$\sim CH^* - CH_2 - [Cat^*] + CH_2 = CH \longrightarrow \sim CH^* - CH_2 - [Cat^*]$$

with pendant groups:

$\overset{|}{C}H$, $\overset{|}{C}H_2$, $H - \overset{*}{C} - CH_3$, C_2H_5

and

$\overset{|}{C}H_2$, $H - \overset{*}{C} - CH_3$, C_2H_5

$$\longrightarrow \sim CH^* - CH_2 - [Cat^*]$$

$\overset{|}{C}H_2$, $H - \overset{*}{C} - CH_3$, C_2H_5

$\overset{|}{C}H_2$, $\overset{|}{C}H$, $H - \overset{*}{C} - CH_3$, C_2H_5

$$\longrightarrow \sim CH^* - CH_2 - \overset{*}{C}H - CH_2 - [Cat^*]$$

$\overset{|}{C}H_2$, $H - \overset{*}{C} - CH_3$, C_2H_5

$\overset{|}{C}H_2$, $H - \overset{*}{C} - CH_3$, C_2H_5

$$\sim CH^* - CH_2 - [Cat^*]$$

$\overset{|}{C}H_2$, $\overset{|}{C}H$, $CH^* - CH_3$, C_2H_5, $H - \overset{*}{C} - CH_3$, C_2H_5

29

Assuming for an isotactic macromolecule of poly-(S)-4-methyl-1-hexene a molar optical activity per monomer unit [Φ]+300 (13) and a linear relationship between rotatory power and excess of monomeric units of a single chirality, the corresponding specific optical activity for the above mentioned statistical copolymer should be about 63, 21, 7.5, or 6. For an atactic polymer of the (S) antipode, even if a quantitative separation can be achieved, an average optical activity of only 38, 12.5, 4.5, or 3.6 respectively, should be found, as the atactic poly-(S)4-MH has [Φ]$_D^{25}$+ 180 (14). If in a portion of a polymer from the racemic monomer an optical activity is found larger than that calculated for a random (R) (S) copolymer a certain degree of stereoselectivity in the polymerization process can be assumed(3).

Attempts to separate poly-α-olefins obtained from racemic monomers by chromatographic separation have been successful in some cases. For instance a separation was indeed achieved, both by solvent extraction of poly-(R)(S)4-MH supported on a crystalline insoluble optically active poly-α-olefin (as poly-(S)-3-methyl-1-pentene), and by elution chromatography on the same support(2,6) as shown in Fig.1. The separation into fractions having opposite optical activity is accompanied by separation according to molecular weight and stereoregularity(6). The experimental separation degree in fractions having opposite optical rotation (F) can be evaluated by eq. 1 (6)

$$F = \frac{2.10^2}{W_t} \sum_{i=1}^{m} \frac{[\alpha]_i}{[\alpha]_o} W_i \qquad [1]$$

where W_t is the total weight of eluted polymer, m is the number of fractions having optical activity of the same sign, W_i is the weight of the i-th fraction having optical activity $[\alpha]_i$, and $[\alpha]_o$ is the optical activity of the corresponding polymer derived from a pure antipode. However, the relation between $[\alpha]$ and enantiomeric purity is in some cases not linear(13); therefore for a more accurate evaluation, the enantiomeric purity should be derived from the relationship between $[\alpha]$ and polymerized monomer optical purity(13). In general the optical rotation observed in the separated fractions is small, the variations of F thus obtained are not very large and we discuss here the values calculated according to eq. 1. These values do not permit to discriminate between the two hypotheses given in Scheme 1. However, under similar separation conditions, the poly-(R)(S)-4-methyl-1-hexene obtained by hydrogenation of the polymer of racemic 4-methyl-1-hexene(14) was not separable into fractions with optical rotations of opposite sign (Scheme 3). As the soluble catalyst

30

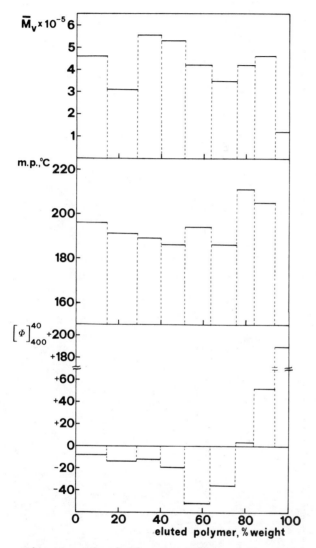

Fig. 1. Physical properties of the fractions obtained by chromatography of diethylether sol., ethylacetate insol. poly-(R)(S)-4-methyl-1-hexene on poly-(S)-3-methyl-1-pentene. (See Ref. 6).

SCHEME 3.

Preparation of poly-(R)(S)-4-methyl-1-hexene by hydrogenation of the polymer from racemic 4-methyl-1-hexene (14).

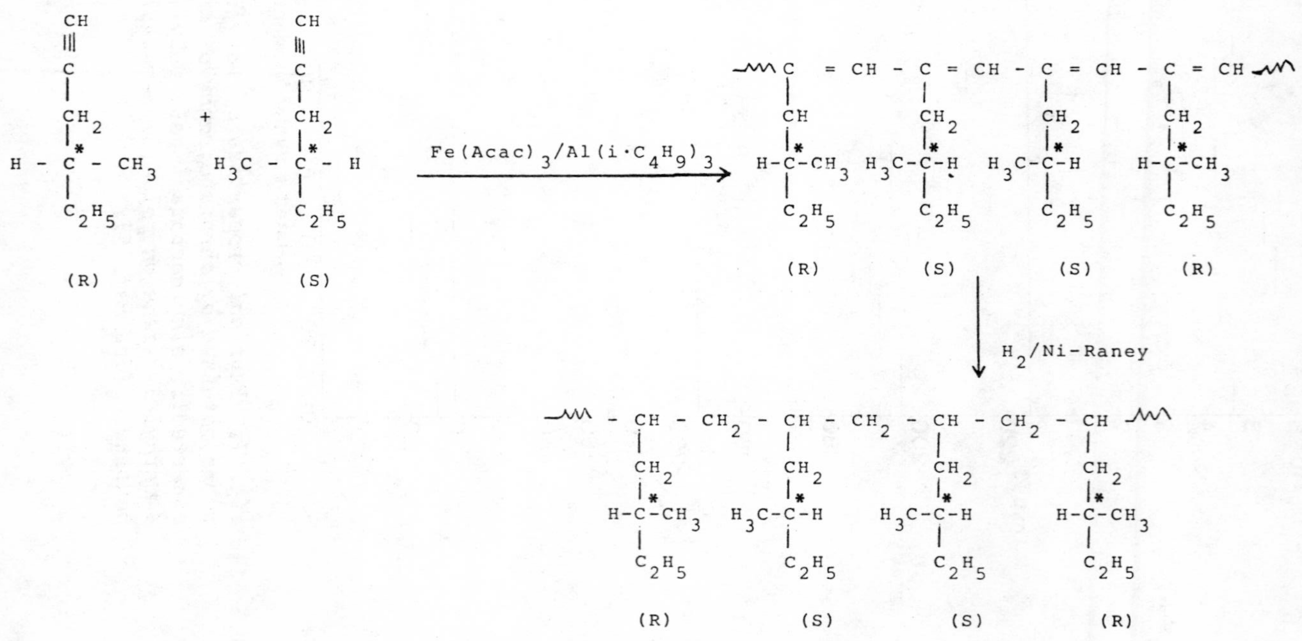

32

Fe(AcAc)$_3$/AlR$_3$ used for the polymerization of 4-methyl-1-hexene is probably not stereoselective, so the lack of separation confirms that a statistical (R)(S)-copolymer is not separable into fractions having opposite optical activity, at least under the conditions used. Therefore, the separation obtained in the case of poly-(R)(S)-4-methyl-1-hexene can be taken as an indication of the existence of stereoselectivity. Further confirmation is obtained from the quantitative estimates of stereoselectivity. To accomplish this it was necessary to know the efficiency E of the separation method which is related to the stereoselectivity degree or "intrinsic separability"[+], D, by the eq. 2.

$$E = F/D \qquad [2]$$

E was evaluated in the following way(15): (R) and (S)-4-MH having comparable optical purity ($\sim$87%) were polymerized separately with TiCl$_4$-Al(i-C$_4$H$_9$)$_3$ catalyst under similar conditions. Moderately crystalline fractions extracted from the homopolymer of each antipode have been mixed in equimolar amounts to give an optically inactive mixture. This last was separated into fractions having optical activity of opposite sign by chromatography on poly-(S)-3-methyl-1-pentene. The same conditions were used in the separation of the polymer derived from the corresponding racemic monomer. As D for the homopolymers mixture can be assumed to be 100%, the evaluation of F directly gives E according to eq. 2. The results are summarized in Scheme 4. The value of E thus obtained for 4-MH is 32.5% and the stereoselectivity ranges from 15 to 20%

[+] By assuming that stereoselectivity coincides with D, we neglect the possible contribution to the separation in the random copolymer. This last should be in any case rather small or even zero as shown in the case of hydrogenated poly-(R)(S)-4-methyl-1-hexene. Some attempts(6) have been made to improve the efficiency of separation using different supports: with poly-(R)-3,7-dimethyl-1-octene as support the first eluted fractions of poly-4-MH have chirality opposite to the one observed using poly-(S)-3-methyl-1-pentene. These experiments show that the mechanism of separation is rather complicated and that the factors regulating the separation of the polymers into fractions having optical activity of opposite sign are far from being understood. In conclusion, the data of Table 1 show that polymerization of racemic 4-MH with conventional Ziegler-Natta catalysts is partially stereoselective. Stereoselectivity seems doubtful only for acetone soluble, atactic and low DP fractions.

SCHEME 4.

Evaluation of the efficiency E of the separation of polymers from racemic α-olefins in fractions having optical activity of opposite sign.

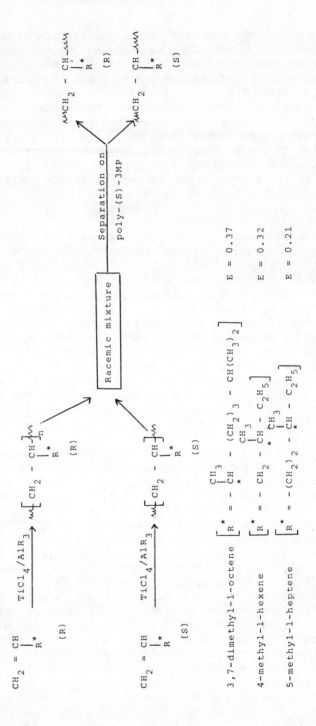

3,7-dimethyl-1-octene $\left[R^* = -CH - (CH_2)_3 - CH(CH_3)_2 \right]$ E = 0.37

4-methyl-1-hexene $\left[R^* = -CH_2 - CH - C_2H_5 \right]$ E = 0.32

5-methyl-1-heptene $\left[R^* = -(CH_2)_2 - CH - C_2H_5 \right]$ E = 0.21

for moderately or highly crystalline fractions. These values demonstrate that the process is stereoselective because D is higher than that calculated for a statistical copolymer of the two antipodes (Table 1).

2.1) Influence of Monomer Structure

In addition to 4-MH other olefins were investigated in order to relate stereoselectivity to monomer structure. As expected, stereoselectivity is strongly influenced by the distance of the asymmetric carbon atom from the double bond (Table 2). Very high stereoselectivity (larger than 90%) is observed for racemic 3,7-dimethyl-1-octene [(R)(S)-3,7-DMO] (15), where the asymmetric carbon atom is in the α-position with respect to the double bond. In the separation process the polymer obtained from racemic samples of this monomer behaves practically as an equimolar mixture of the homopolymers of the two antipodes. No stereoselectivity occurs for racemic 5-methyl-1-heptene [(R)(S)-5-MH] in which the asymmetric carbon atom is separated from the double bond by two CH$_2$ groups, indicating that when the chiral center is in the γ position, the catalytic centers are not able to distinguish between the two monomer antipodes(17).

2.2) Influence of Catalyst Nature

A further point of interest was to establish if stereoselectivity exists only for a conventional heterogeneous Ziegler-Natta catalyst, or if it is more general phenomenon. As shown in Table 3 stereoselectivity for (R)(S)-4-MH seems doubtful only in the case of soluble catalyst obtained from Ti(CH$_2$C$_6$H$_5$)$_4$ and Al(CH$_2$C$_6$H$_5$)$_3$ (18) which also are not very stereospecific. Conventional catalysts show similar stereoselectivity(6), while a much higher value has been achieved more recently in the case of racemic 4-MH using a catalyst obtained from AlR$_3$ and TiCl$_4$ supported on MgCl$_2$ (18).

2.3) Stereoselectivity in the Copolymerization of Racemic α-Olefins with Ethylene

Racemic 4-MH (19) and 3,7-DMO (20,21) have been copolymerized with ethylene in the presence of conventional heterogeneous Ziegler-Natta catalysts into copolymers having substantially statistical distribution of the two comonomers, as shown by fractional precipitation of copolymerization products. These have been submitted to chromatographic separation on poly-(S)-3-methyl-1-pentene under the same conditions used

TABLE 1.

Separation degree (F) and intrinsic separability (D) of polymer samples from racemic 4-methyl-1-hexene [(R)(S)4-MH] and of equimolar mixtures of poly-(R)4-MH and poly-(S)4-MH.

Sample submitted[a] to chromatographic separation		$\overline{DP}_v$	$\overline{DP}_n$	Separation degree, (F) %	Intrinsic[b] separability (D) %	Theoretical separability of a statistical copolymer of the two antipodes[c] %
Atactic poly-(R)(S)4-MH[d]		(1,800)[e]		0	--	2
Poly-(R)(S)4-MH	1		10	2.5	7.7	26
	2		100	4.9	15.0	7
	3	(4,600)[e]		5.1	16.0	>1, <2
	4	(2,100)[e]		5.5	17.0	2
	5	(3,900)[e]		6.1	19.0	>1, <2
Poly-(S)4-MH and poly-(R)4-MH equimolar mixture	2	100	100	32.5	100	---

[a] Extracted successively with: 1) acetone; 2) ethylacetate; 3) diethylether; 4) diisopropylether; 5) isooctane. [b] $\underline{D} = F/\underline{E}$, where $\underline{E}$ is the efficiency of the separation. [c] For a monodisperse polymer with DP equal to $\overline{DP}_n$ of the separated sample. [d] Obtained by hydrogenation of poly-(R)(S)-4-methyl-1-hexyne (14). [e] Evaluated from intrinsic viscosity by the equation $[\eta] = 3.51 \cdot 10^{-5} \overline{M}_v^{0.83}$ determined for poly-(R)(S)4-MH (16).

TABLE 2.

Influence of the monomer structure on the stereoselectivity of polymers from racemic α-olefins.

Polymer submitted to chromatographic separation		$\overline{DP}_v$ [b]	$\overline{DP}_n$	Separation degree, % (F)	Separability % (D) [c]	Theoretical[d] separability %, of a statistical co-polymer of the two antipodes (D_o)
Type	Fraction					
Poly-(R)(S)3,7-DMO	Ethyl acetate ins. diethyl ether sol.	(2,700)[b]		33.6	91	>1, <2
Poly-(R)3,7-DMO Poly-(S)3,7-DMO[a]	Acetone ins. ethyl acetate sol.		45	37.0	100	--
Poly-(R)(S)4-MH	Diisopropylether ins. isooctane sol.	(3,900)[b]		6.1	19	>1, <2
Poly-(R)4-MH Poly-(S)4-MH[a]	Acetone ins. ethyl acetate sol.		100	32.5	100	--
Poly-(R)(S)5-MH		(3,400)[b]		0	--	>1, <2
Poly-(R)5-MH Poly-(S)5-MH[a]	Methanol ins. diethylether sol.	(4,000)[b]		21	100	>1, <2

[a] Equimolar mixture. [b] Evaluated from intrinsic viscosity using the eq. $[\eta] = 3.51 \cdot 10^{-5} (\overline{M}_v)^{0.83}$, reported for poly-4MH (16). [c] $\underline{D} = F/\underline{E}$ where $\underline{E}$ is the efficiency of the separation. [d] For a monodisperse polymer with DP equal to $\overline{DP}_n$ of the separated sample.

TABLE 3.

Influence of the type of catalytic system on the stereoselectivity in the polymerization of racemic 4-methyl-1-hexene [(R)(S)4-MH].

Catalysts's components		Fraction	Maximum separation[a] degree achieved (F)	Ref.
Metal halide (MX_n)	Metal organic (MeRm)			
$TiCl_3$ "ARA"	$Al(\underline{i}\text{-}C_4H_9)_3$	Diethyl ether soluble ethyl acetate insol.	5.1	6
$TiCl_3 \cdot N(n\text{-}C_4H_9)_3$	---	Benzene soluble[b] diethylether insol.	3.5[b]	18
$Ti(CH_2C_6H_5)_4$	$Al(CH_2C_6H_5)_3$	Diethylether soluble ethyl acetate insol.	0.6	
$TiCl_4/MgCl_2$	$AlEt_3$[d]	Diethylether sol., ethyl acetate insol.	4.3	18
		Benzene soluble diethylether insol.	15.4[c]	

[a] By elution chromatography on poly-(S)-3-methyl-1-pentene if not otherwise indicated.
[b] Sample kindly supplied by Dr. J. Boor, Jr. [c] By solvent extraction on poly-(S)-3-methyl-1-pentene. [d] In the presence of esters of aromatic acids.

38

for the corresponding racemic monomers. Both in the case of a 4-MH/ethylene copolymer containing 22% in mole of the latter and a 3,7-DMO/ethylene copolymer with 44% in mole of ethylene units, the separation into fractions having optical activity of opposite sign was achieved. The separation degree was calculated by taking into account the optical rotation of copolymers of ethylene with optically active 4-MH and 3,7-DMO having chemical composition similar to that of the separated copolymer. The separation degree is very close to that previously observed for the homopolymers of racemic 4-MH and 3,7-DMO (Table 4). Taking into consideration the probably lower efficiency of the support in the separation of the copolymers with respect to homopolymers, it can be concluded that the insertion of ethylene units in the growing chains neither cancels stereoselectivity of the polymerization nor decreases it appreciably.

3) *FURTHER EXPERIMENTS IN THE FIELD OF RACEMIC α–OLEFIN POLYMERIZATION AND COPOLYMERIZATION*

As it will be discussed later in details, (see Sect. 4) the existence of stereoselectivity in non-sterospecific polymerization of racemic 4-MH and in copolymerization of racemic α–olefins with ethylene indicated that the existence of asymmetric carbon atoms in the growing chain does not play an appreciable role, if any, in determining stereoselectivity of α–olefins polymerization. The ability of the catalytic centers to distinguish between the two antipodes of a monomer, as required for stereoselective polymerization, must therefore arise from their intrinsic asymmetry (for instance from the existence of chiral titanium atoms). The above results stimulated further research on racemic α–olefin polymerization and copolymerization. Firstly, attempts have been made to modify the catalytic centers in order to polymerize preferentially one of the antipodes of the monomer(22). Subsequently, copolymerization of a racemic monomer with an optically active comonomer having similar structure, was attempted in order to show that steric factors can have more importance than chemical structure in copolymerization by Ziegler-Natta catalytic systems(23). Finally, monomers having different optical purity were polymerized in order to better understand the relationship between the prevalence of monomeric units arising from an antipode and the optical activity of the resulting polymers(24). These three series of experiments which, besides confirming the existence of stereoselectivity, have given further indications concerning Ziegler-Natta catalysts, will be discussed in the following sections.

TABLE 4.

Separation degree (F) in fractions having opposite optical activity for homopolymers of racemic 3,7-dimethyl-1-octene [(R)(S)-3,7-DMO] and 4-methyl-1-hexene [(R)(S)-4-MH)] and for their copolymers with ethylene

Polymer separated on poly-(S)-3-methyl-1-pentene	Fraction	Separation degree F (%)	Ref.
poly-(R)(S)4-MH	Diethylether soluble, ethylacetate insol.	5.8	19
(R)(S)4-MH/C_2H_4 (78/22) copolymer	Diethylether soluble, ethylacetate insol.	4.2 - 7.0	
poly-(R)(S)3,7-DMO	Diethylether soluble, acetone insoluble	27.3	21
	Diethylether soluble, ethylacetate insol.	33.6	
(R)(S)3,7-DMO/C_2H_4 (56/44) copolymer	Cyclohexane soluble, acetone insoluble	23 - 29	

40

3.1) Stereoelective Polymerization

According to the results of stereoselective polymerization, Ziegler-Natta catalysts can be considered to be a racemic mixture of catalytic centers having opposite chirality, even if the asymmetric carbon atoms of the growing chains are ignored. Therefore, even if the overall polymerization rate of the two antipodes must be the same, the polymerization rate of one antipode at a single chiral catalytic center must be largely different from that of its enantiomer. In order to polymerize electively only one antipode, centers of a single chirality must be prepared. If the catalytic centers contain more than one chiral center (for instance a chiral metal atom and a chiral group bound to it) diastereomeric catalytic centers are produced from a racemic mixture of catalytic centers and an optically active ligand, a stereoelective catalyst can be produced. As the nature of the catalytic centers is not known, a rational approach to this problem is presently impossible. As the catalytic centers are in general formed by reaction of a transition metal compound with a metal alkyl, the most simple empirical approach to produce stereoelective catalysts is an attempt to synthesize diastereomeric catalytic centers by reacting either a transition metal halide with an optically active metal alkyl, or a transition metal compound containing optically active ligands with non-chiral metal alkyls. This approach has allowed us to achieve both the first asymmetric polymerization of a non-chiral diolefin (25) and the stereoelective polymerization of the racemic α-olefins(21) (Scheme 5).

The first attempts(26) to prepare a stereoelective catalyst for α-olefins polymerization was made using $TiCl_4$ and tris-[(S)-2-methyl-butyl]-aluminum diethyletherate. In this case the (S)-2-methylbutyl groups were found (Scheme 6) as expected(27) as terminal groups in the polymerization products both of prochiral (propylene and styrene) and of racemic α-olefins(26) (RS-4-MH). However, no detectable optical activity was found in the non-polymerized recovered monomers when starting with racemic 4-MH while optical activity could be expected in the case of stereoelective polymerization of a racemic monomer (Scheme 5). Therefore, we can conclude that using the above catalytic system and monomer, the presence of an optically active terminal group is not sufficient to increase appreciably the polymerization rate of one of the monomer antipodes with respect to that of its enantiomer even if the polymerization is stereoselective. The situation is different in the case of catalysts prepared from $TiCl_4$ and bis-[(S)-2-methylbutyl]-zinc. In this case the monomer anti-

SCHEME 5.

Synthesis of optically active polymers by optically active Ziegler-Natta catalysts.

a.) Asymmetric synthesis of cis-1-4-isotactic-polypentadiene 25)

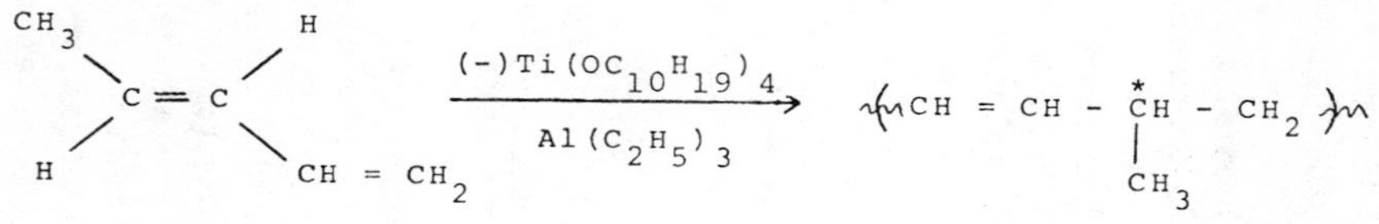

$$CH_3\text{—}C=C\text{—}CH=CH_2 \xrightarrow[Al(C_2H_5)_3]{(-)Ti(OC_{10}H_{19})_4} \{CH=CH-\overset{*}{CH}-CH_2\}_n$$

with H substituents and CH_3 on the starred carbon.

b.) Stereoelective polymerization of racemic α-olefins 22)

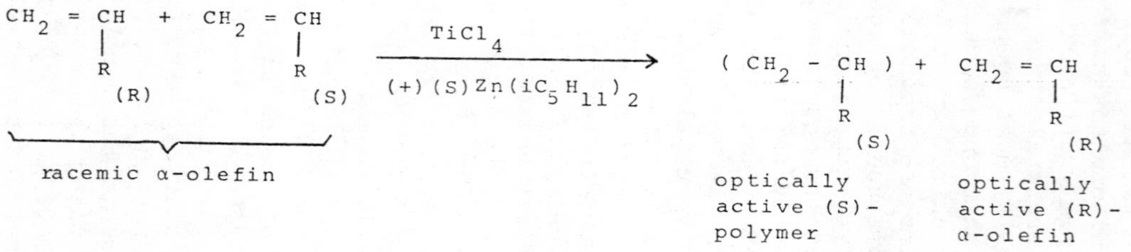

$$\underbrace{CH_2=CH + CH_2=CH}_{\text{racemic α-olefin}} \xrightarrow[(+)(S)Zn(iC_5H_{11})_2]{TiCl_4} (CH_2-CH) + CH_2=CH$$

with R (R) and R (S) substituents;

optically active (S)-polymer   optically active (R)-α-olefin

SCHEME 6.

Growing of a macromolecular chain on a catalytic site obtained from transition metal halides and optically active metalalkyls (26).

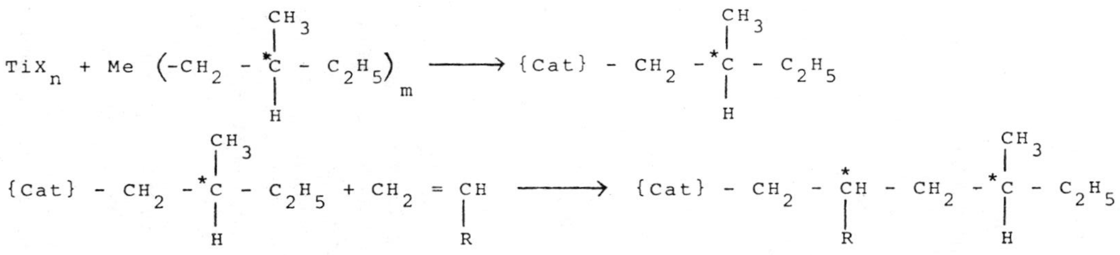

$$TiX_n + Me \left(-CH_2 - \overset{*}{\underset{\underset{H}{|}}{\overset{\overset{CH_3}{|}}{C}}} - C_2H_5 \right)_m \longrightarrow \{Cat\} - CH_2 - \overset{*}{\underset{\underset{H}{|}}{\overset{\overset{CH_3}{|}}{C}}} - C_2H_5$$

$$\{Cat\} - CH_2 - \overset{*}{\underset{\underset{H}{|}}{\overset{\overset{CH_3}{|}}{C}}} - C_2H_5 + CH_2 = \underset{\underset{R}{|}}{CH} \longrightarrow \{Cat\} - CH_2 - \overset{*}{\underset{\underset{R}{|}}{CH}} - CH_2 - \overset{*}{\underset{\underset{H}{|}}{\overset{\overset{CH_3}{|}}{C}}} - C_2H_5$$

$$R = -CH_3, \quad -C_6H_5$$

pode having the same chirality as the optically active metal-organic compound, was preferentially polymerized in the case of racemic 3-methyl-1-pentene [(R)(S)-3-MP](22,28), 3,4-dimethyl-1-pentene(29) [(R)(S)-3,4-DMP], 4-MH(30) and 3,7-DMO (22,28), while no stereoelectivity was found for 5-MH (Table 5).

A systematic investigation of the stereoelective polymerization was carried out and the results can be summarized as follows:

(1) Using $TiCl_4$ as the transition metal component, the stereoelectivity depends on the type of metal alkyl compound used (Table 6); with the same asymmetric groups bound to the metal, the stereoelectivity decreases according to the following order: $ZnR_2 >> AlR_3 \cdot Et_2O > AlR_3 \sim InR_3 > GaR_3 > BeR_2 = LiR$ (30).

(2) Using ZnR_2 the type of group bound to Zn has an influence on stereoelectivity with: $-CH_2-CH(C_6H_5)-CH_3 > -CH_2-CH(C_2H_5)-CH_3 > -CH_2-CH(C_6H_5)-iC_3H_7$.

(3) Using bis[(S)-methylbutyl]-zinc, the stereoelectivity depends on the type of metal halide used(29): $TiCl_4 > TiBr_4 > TiI_4 > VCl_4 > ZrCl_4 > NbCl_5$.

(4) A stereoelective catalyst can be obtained using an optically active titanium compound (for instance *I*) and tris-(benzyl)aluminum(31).

$$CH_3-\underset{\underset{\displaystyle CH_3-\overset{*}{CH}-O}{\underset{\displaystyle CH_2}{|}}}{\overset{\overset{\displaystyle CH_3}{|}}{C}}-O \diagdown Ti(CH_2-C_6H_5)_2$$

(*I*)

(5) A stereoelective polymerization catalyst can be obtained by adding to a conventional catalyst $(TiCl_4/AlR_3)$, an optically active electron donor, as internal olefins, aromatic hydrocarbons, or amines(32).

(6) Stereoelectivity decreases by increasing the distance between the asymmetric carbon atom and the double bond in the olefin. No stereoelectivity is observed when the double bond is in the γ position with respect to the asymmetric carbon atom(30).

(7) No effect on stereoelectivity has been observed by varying the conversion of monomers into polymers from 2% to 20% (29) (Table 7).

In conclusion, stereoelectivity is in general much lower than stereoselectivity showing that the modification of enantiomeric catalytic sites, due to the use of optically active

TABLE 5.

Stereoelective polymerization of racemic α-olefins by $TiCl_4$/bis-[(S)-2-methylbutyl]-zinc catalytic system.

| Monomer | Conversion, (C, %) | Non-polymerized monomer | | Optical rotation of the last extracted fraction ($[\alpha]_D^{25}$) | Polymerized monomer optical purity $(P_p)^a$ (%) | Ref. |
		$[\alpha]_D^{25}$	Optical purity (Pm) (%)			
(R)(S)-3-MP	7.2	-0.75[b]	2.0	+14.5[c]	20.0	22
(R)(S)-3,4-DMP	15.0	-0.39[b]	0.9	+15.5	5.0	29
(R)(S)-3,7-DMO	26.9	-0.63[b]	3.9	+16.1	10.5	28
(R)(S)-4-MH	19.0	+0.07[d]	0.7	+ 1.9	2.6	29
(R)(S)-5-MH	24.3	0.0	0.0	+ 1.6[f]	0.0	30

[a] P_p = P_m (100-c)/c. [b] λ = 589 nm. [c] At 60°C. [d] λ = 365 nm. [e] λ = 300 nm. [f] $\overline{M}_n$ = 2,000.

TABLE 7.

Variation of stereoelectivity versus conversion in the polymerization of racemic 3,7-dimethyl-1-octene [(R)(S)-3,7-DMO] with optically active metal alkyls (MeR$_n$) and TiCl$_4$ catalyst(29).*

MeR*$_n$	Conversion (%)	Non-polymerized monomer $[\alpha]^{25}_D$	Optical purity (P_m), (%)	Polymerized monomer optical purity (P_p)[a] (%)	Polymer optical activity Methanol soluble fraction	Methanol insoluble fraction
Bis[(S)-2-methyl-butyl]-zinc	3.5	-0.062	0.38	3.2	--	--
	12.1	-0.297	1.82	4.0	+5.4	+13.5
	18.4	-0.497	3.05	3.9	+5.1	+10.4
	28.6	-1.14	6.99	5.3	+4.3	+11.2
Tris-[(S)-2-methylbutyl]-aluminum	2.3	-0.007	0.04	0.12	+2.4	+ 0.9
	4.8	-0.034	0.21	0.20	+2.2	+ 0.6
	17.5	-0.150	0.92	0.33	+1.3	+ 0.5

[a]See Table 5.

TABLE 6.

Stereoelective polymerization of racemic 3,7-dimethyl-1-octene [(R)(S)-3,7-DMO] by TiCl₄ and different optically active metalalkyls (MeR*ₙ)(30).

MeR*ₙ	Conversion %	Non-polymerized monomer		Optical rotation of the last extracted polymer fraction ($[\alpha]_D^{25}$)	Polymerized monomer optical purity (P_p),[a] (%)
		$[\alpha]_D^{25}$	Optical Purity (P_m), %		
LiR*	30.9	0	0	0.0	0
BeR*₂	30.2	0	0	0.0	0
AlR*₃·Et₂O	7.9	−0.024	0.15	+2.0	1.7
AlR*₃	35.2	−0.57	0.62	+1.0	1.1
GaR*₃	31.0	−0.19	0.21	+0.5	0.5
InR*₃	35.2	−0.63	0.69	+0.8	1.3
ZnR*₂	26.9	−0.63	3.86	+16.1	10.5

[a] P_p, see Table 5.

transition metal or organometallic compounds, causes only a
rather small difference in the polymerization rate of the two
antipodes. Of the two possible explanations for the above
difference, i.e. a different number of catalytic centers
having opposite chirality, or a modification of the reactivity
of the two enantiomeric sites towards the two antipodes of the
monomer by introducing   a new asymmetric center having a
predominant chirality, the second appears more likely. However, the nature of the modification of the catalytic centers
created by chiral catalyst components is still obscure,
although some attempts have been made to correlate the difference in stereoelectivity of catalytic systems prepared
from zinc alkyls and aluminum alkyls with the possible modification of the catalytic centers. No sound general explanation is at present available for stereoelectivity found in
polymerization of racemic monomers with optically active
catalysts, in polymerization of monomers having different
optical purity (Sect. 3.3) and in copolymerization of racemic
with optically active monomers (Sect. 3.2).

3.2) Copolymerization of Racemic α-Olefin with an Optically Active α-Olefin

    Due to the stereoselective character of the polymeriza-
tion of α-olefins by Ziegler-Natta catalyst, the copolymerization of an optically active α-olefin(*A*) with a racemic α-
olefin(*B*) should yield a copolymer of *A* with the antipode of
B, having the same chirality and homopolymer of the remaining
antipode of *B* (Scheme 7). The occurrence of this stereoselective copolymerization can be easily demonstrated by
solvent extraction of the polymerization product. In fact,
the copolymer of two monomers each of which gives a crystalline homopolymer, is more soluble than the two corresponding
homopolymers and can be extracted with relatively poor solvents. This result was indeed obtained by copolymerizing
racemic 3,7-DMO with (S)-3-MP as well as racemic 3-MP with
(R)-3,7-DMO in the presence of $TiCl_4$-AlR_3 (or ZnR_2) catalyst
(23,34). In the former case the copolymer consisting of S-
units must have positive optical rotation(32) while the opposite holds for the homopolymer(35). Accordingly, the first
fractions extracted with relatively poor solvents have positive and the last fraction has negative optical activity.
Moreover, IR analysis indicates that the amount of 3-MP units
decreases and that of 3,7-DMO units increases in the fractions
extracted successively (Table 8). Analogous results were
obtained for the copolymerization of (R)(S)-3-MP with (R)-3,7-
DMO. These data prove beyond any doubt that a stereoselective

SCHEME 7.

Stereoselective copolymerization of a racemic α-olefin (B) with an optically active α-olefin (A) (34).

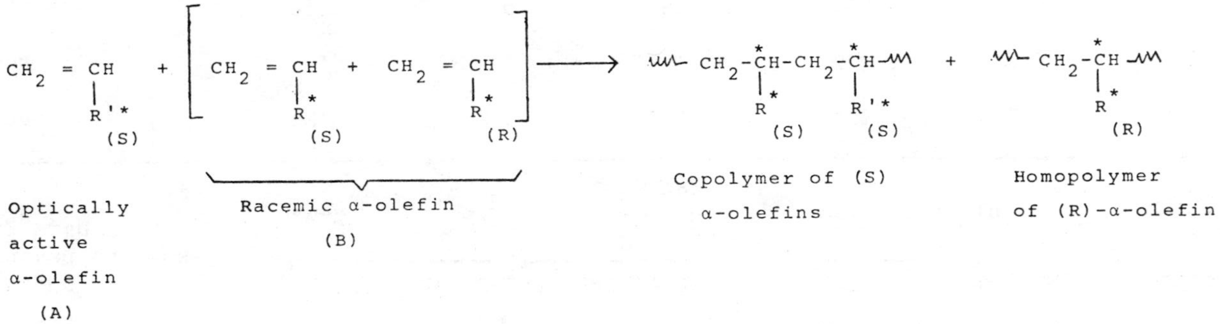

$$CH_2 = CH \atop R'^* \quad (S)} \quad + \quad \left[\begin{array}{c} CH_2 = CH \\ R^* \\ (S) \end{array} + \begin{array}{c} CH_2 = CH \\ R^* \\ (R) \end{array} \right] \longrightarrow$$

Optically
active
α-olefin

(A)

Racemic α-olefin

(B)

$$\sim\!\!\!\wedge\!\!\!\wedge CH_2 - \overset{*}{C}H - CH_2 - \overset{*}{C}H \sim\!\!\!\wedge\!\!\!\wedge \atop R^* \quad\quad R'^* \atop (S) \quad\quad (S)} \quad + \quad \sim\!\!\!\wedge\!\!\!\wedge - CH_2 - \overset{*}{C}H \sim\!\!\!\wedge\!\!\!\wedge \atop R^* \atop (R)}$$

Copolymer of (S)
α-olefins

Homopolymer
of (R)-α-olefin

49

TABLE 8.

Physical properties of fractions obtained by extracting with boiling solvents the polymeric products prepared by copolymerizing (R)(S)-3,7-dimethyl-1-octene [(R)(S)-3,7-DMO] with (S)-3-methyl-1-pentene [(S)-3-MP].

Fractions extracted successively with	TiCl$_4$/Zn(i-C$_4$H$_9$)$_2$ catalyst			TiCl$_4$/Al(i-C$_4$H$_9$)$_3$ catalyst		
	wt-%	$[\alpha]_D^{25}$	$\dfrac{D_B\ 763^{a}}{D_B\ 732}$	wt-%	$[\alpha]_D^{25}$	$\dfrac{D_B\ 763^{a}}{D_B\ 732}$
Methanol	42.0	+ 3.0	n.d.	14.1	+ 1.2	n.d.
Acetone	16.9	+26.1	n.d.	8.8	+ 5.0	(0.53)
Ethylacetate	29.7	+64.9	0.56	11.6	+49.5	0.43
Diethylether	8.5	+38.0	0.60	37.2	+76.0	0.55
Diisopropylether	2.9	−51.6	0.20	2.7	+28.6	0.44
Cyclohexane	0	——	——	25.6	−19.6	0.18

[a]Copolymer composition. See Ref. 34.

process has taken place giving a (S)-3-MP/(S)-3,7-DMO (or (R)-3-MP/(R)-3,7-DMO) copolymer and poly-(R)-3,7-DMO (or poly-(S)-3-MP). The method of separation used, based on solvent extraction is very simple, but its efficiency is probably not very high. Therefore a more accurate analysis of polymer composition and a larger number of examples seem to be necessary for a better understanding of the influence of catalyst and monomer structure. An improvement of separation efficiency would be necessary to develop the potential usefulness of the process for preparation of optical active polymers from a racemic monomer and a small amount of a suitable optically active comonomer.*

In all cases investigated up to now, the nonpolymerized monomer was optically active, the sign of the optical rotation indicating the prevalence of the antipode having opposite configuration to A. Therefore, the stereoselective copolymerization is accompanied by a stereoelective process, preferentially including in the polymer chains units derived from the antipode of B having the same chirality as the optically active comonomer A (Table 9). In order to get information about this phenomenon, (R)(S)-3,7-DMO was polymerized in the presence of catalysts obtained by reacting TiCl₄ with bis [(S)-3-methylpentyl]-zinc(23) or tris-[(S)-3-methylpentyl]-aluminum(37); the ratio (R)(S)-3,7-DMO/(S)-3-methylpentyl group was equal to the (R)(S)-3,7-DMO/(S)-3,7-DMO/(S)-3-MP ratio used in the copolymerization experiments. In the former case a stereoelectivity was obtained very close to that obtained with Zn(i-C₄H₉)₂ and (S)-3-MP, while tris-[(S)-3-methylpentyl]-aluminum gave no stereoelectivity. These data, even if they do not allow us to depict a simple mechanism for explaining the stereoelective character of the polymerization, exclude the possibility that with Al(i-C₄H₉)₃ and (S)-3-MP, the stereoelectivity is due to the formation of tris-[(S)-3-methylpentyl]-aluminum in the reaction mixture.

3.3) Polymerization of α-Olefins having Different Optical Purity.

In the case of low molecular weight compounds the optical rotation depends linearly on the optical purity. A linear dependence of rotatory power vs. optical purity of polymerized

---

* The method has been successfully applied in the case of chiral vinylethers in the presence of heterogeneous stereospecific catalyst not containing transition metals(36).

TABLE 9.

Stereoelectivity in the polymerization of a racemic α-olefin (B) in the presence of an optically active α-olefin (A) (34).

Comonomers		Catalyst	Conversion[a] C (%)	Optical purity of B	
Type	Molar ratio (B/A)			Non-polimerized P_m (%)	Polimerized P_p (%)[b]
(R)(S)-3,7-DMO (S)-3-MP[c]	3.41	$TiCl_4$–$Zn(i$-$C_4H_9)_2$	21.2	1.2	4.4
(R)(S)-3,7-DMO (S)-3,7-DMO[c]	26.0	$TiCl_4$–$Zn(i$-$C_4H_9)_2$	34.3	0.9	1.7
(R)(S)-3-MP (S)-3-MP[c]	3.36	$TiCl_4$–$Al(i$-$C_4H_9)_3$	30.1	0.8	1.9
(R)(S)-3-MP (R)-3,7-DMO[d]	3.36	$TiCl_4$–$Zn(i$-$C_4H_9)_2$	24.4	1.7	5.4

[a](Weight of total polymer/weight of starting comonomers mixture) 100. [b]See Table 5.

[c]Optical purity 89%. [d]Optical purity 95%.

monomer could be also expected if stereoselectivity is 100%, or if it does not occur. The optical activity of the whole polymer from samples of (R)-3,7-DMO or (S)-4-MH having different optical purity, does not follow a straight line, while a linear relationship is observed for polymers from (S)-5-MH(13) (Fig. 2). The rotatory power is in fact higher than of the corresponding homopolymers mixture for each composition; this could be due either to the preferential polymerization of the predominant enantiomer or to conformational effects or to both factors. Further investigation has shown that the second effect is the predominant one. In fact, only in the presence of $TiCl_4/ZnR_2$ catalyst was an appreciable preferential polymerization of the predominant (S) antipode observed, as in the copolymerization of (R)(S)-3,7-DMO with (S)-3-MP (Table 10)(24).

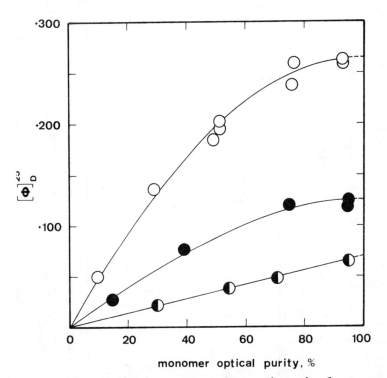

Fig. 2. Molar rotatory power versus polymerized monomer optical purity of some poly-(S)-α-olefins: (○) poly-4-methyl-1-hexene; (●) poly-3,7-dimethyl-1-octene; (◐) poly-5-methyl-1-pentene (See Ref. 13).

TABLE 10.

Stereoelectivity in the copolymerization of optically active 3,7-dimethyl-1-octene having different optical purity and in the copolymerization of racemic-3,7-dimethyl-1-octene with (R)-3-methyl-1-pentene. [a]

| Starting monomers mixture | | | | Relative polymerization rate constant of the 2 antipodes of 3,7-DMO (R_p) [b] | Ref. |
| 3,7-DMO | | 3-MP [c] | mole of R monomers / mole of S monomers | | |
R (mole-%)	S (mole-%)	R (mole-%)			
48.15	48.15	3.7	1.07	1.04	34
57.45	42.55	0	1.35	1.07	24
38.65	38.65	22.7	1.51	1.09	34
69.25	30.75	0	2.25	1.19	24
86.75	12.25	0	6.84	1.25	24

[a] $TiCl_4$–$Zn(\underline{i}\text{-}C_4H_9)_2$ catalytic system.

[b] $R_p = \dfrac{\text{[moles of polymerized (R)3,7-DMO/moles of polymerized (S)3,7-DMO]}}{\text{[moles of starting (R)3,7-DMO/moles of starting (S)3,7-DMO]}} = \dfrac{(100+P_p)(100-P_l)}{(100-P_p)(100+P_l)}$,

where P_l = optical purity of starting 3,7-DMO, P_p = optical purity of polymerized 3,7-DMO.

[c] Optical purity 89%.

Also in this case, however, the preferential polymerization
of the (S) antipode was not sufficient to explain the values
of polymer optical rotation. Therefore, the peculiar shape
of the curve optical rotation vs. optical purity in poly-3,7-
DMO and poly-4-MH was interpreted by assuming a conformational
effect, by which the units from the enantiomer present in
larger concentration could induce the units from the other
enantiomer included in the same macromolecule to assume the
allowed conformation(35), giving optical rotation of opposite
sign to that of the corresponding homopolymer. This would
obviously mean that even in the case of 3,7-DMO the stereoselectivity is not complete. A theoretical calculation has
shown(38) that even small deviation from 100% stereoselectivity
could explain the shape of the curve observed. In any case
this could also mean that the stereoselectivity varies with
monomer optical purity, that is, with the relative concentration of the two enantiomers. Examination of polymer fractions
having different stereoregularity obtained from the crude
polymer by solvent extraction shows that the linear dependence
of rotatory power on the optical purity of the polymerized
monomer is observed only for the acetone soluble low molecular weight amorphous fraction. The plots of optical rotation
versus optical purity of all the successively extracted fractions display the same shape, the deviation from linearity
increasing with the degree of stereoregularity. The situation
is somewhat complex for poly-(S)-4-MH, where for optical
purity of the monomer between 30 and 70%, the intermediate
fractions have the same and even larger rotatory power than
the last extracted fraction. As shown by ^{14}C labelled (S)-
4-MH (24), this last fraction contains a larger amount of R
units than the previous ones. This result can be explained,
considering the higher solubility of poly-(S)-4-MH [or poly-
(R)-4-MH] with respect to the equimolar mixture of the two
homopolymers, and can be taken as an indication that stereoselectivity increases with stereospecificity. The last
extracted fraction has in fact, independent of the optical
purity of the starting monomer, a higher IR stereoregularity
index than the other fractions(24). The nonlinear dependence
of the polymer's rotatory power on the polymerized monomer's
optical purity must be considered an additional confirmation
of the stereoselectivity. In fact, if this last were completely lacking, a linear dependence would be expected, as
observed for poly-5-MH. The larger deviation for 4-MH agrees
with the lower stereoselection with respect to 3,7-DMO, where
the asymmetric carbon atom is closer to the double bond.

4) *ORIGIN OF STEREOSELECTIVITY IN THE POLYMERIZATION OF*
    RACEMIC α–OLEFINS

According to the accepted scheme of Ziegler–Natta poly-
merization(27), chain growth occurs by insertion of the monomer molecules between a metal atom of the catalytic center and
the CH_2 group of the ultimate monomeric unit of the growing
chain bound to the catalyst (Scheme 8A). The coupled catalyst-
growing chain is able to distinguish between the two antipodes
of the monomer, the polymerization rate of one antipode at a
single chiral center being in some cases more than 10 times
larger than the polymerization rate of its enantiomer. The
ability to distinguish between two antipodes must arise from
the presence of one or more chiral centers present in the
coupled catalyst-growing chain which interacts with the monomer during its insertion in the growing chain. When a racemic
monomer like 3-MP is polymerized (Scheme 8B), the following
chirality centers can exist in the couple catalyst-growing
chain:

(1) In the growing chain two asymmetric carbon atoms
(a) and (b) exist in the last monomeric unit; furthermore
2(n-1) asymmetric carbon atoms are present in the growing
chain, if n is the number of monomeric units of the growing
chain.

(2) Depending on the type and on the stereochemical
arrangement of the ligands around the metal atom which is
bound to the growing chain, the metal atom (c) can be chiral.
Examples of soluble complexes containing transition(39) or
non-transition(40) metal atoms are well known. When a metal
atom on a solid surface is considered, the conditions necessary
for the existence of chiral atoms can be easily realized(41).

(3) Chirality centers can exist in one or more ligands
(d) bound to the metal.

All the experimental results obtained up to now in the
stereoselective and stereoelective polymerization and copolymerization can be satisfactorily explained excluding any substantial influence by the chiral centers of the growing chain.
In particular, the influence of the chiral centers of the
growing chain (a) can be excluded on the basis of the existence of stereoselectivity in the formation of practically
atactic fractions of poly-(R) (S)-4-MH and of the existence
of stereoselectivity in the copolymerization of racemic α-
olefins with ethylene.

The influence of the asymmetric carbon atoms of the
lateral chains (b) can be excluded on the basis of the copolymerization of racemic olefins with ethylene. In fact, after
the insertion of only one molecule of ethylene, the chiral

SCHEME 8.

Structure of the active catalytic site after insertion of one prochiral monomer (A) or of one chiral monomer molecule (B).

A)

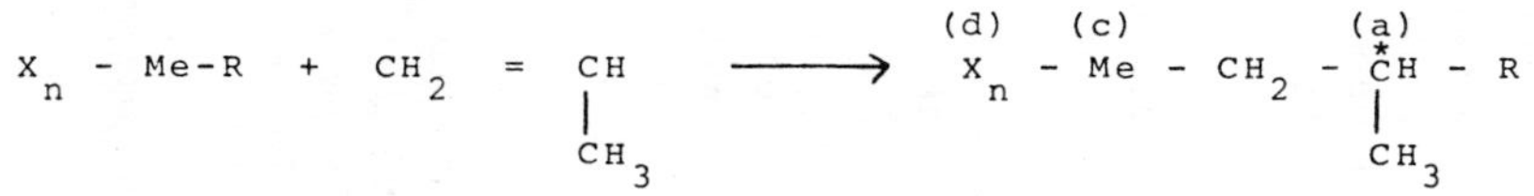

$$X_n - Me-R + CH_2 = \underset{\underset{CH_3}{|}}{CH} \longrightarrow \overset{(d)}{X_n} - \overset{(c)}{Me} - CH_2 - \overset{(a)}{\underset{\underset{CH_3}{|}}{\overset{*}{CH}}} - R$$

B)

$$X_n - Me - R + CH_2 = \underset{\underset{\underset{C_2H_5}{|}}{H - \overset{*}{C} - CH_3}}{CH} \longrightarrow \overset{(d)}{X_n} - \overset{(c)}{Me} - CH_2 - \overset{(a)}{\underset{\underset{\underset{C_2H_5}{|}}{H - \overset{(b)}{\overset{*}{C}} - CH_3}}{\overset{*}{CH}}} - R$$

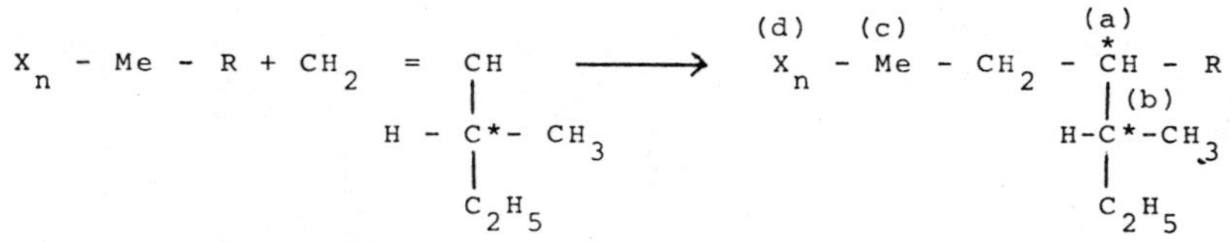

center (b) is in ξ-position with respect to the metal atom of
the catalytic center; therefore no substantial interaction
between the chiral center (b) and the monomer molecule which
is being inserted into the metal-carbon bond of the catalytic
center can be expected (Scheme 9). Other evidence against a
substantial influence of the asymmetric carbon atoms (a) and
(b) on the stereochemistry of racemic α-olefins polymerization arise from the stereoelectivity in the polymerization
and copolymerization processes. In fact: (i) By polymerizing
racemic α-olefins with catalysts prepared from [(S)-2-methyl-
butyl] derivatives of different non-transition metals and
TiCl₄, no definite relationship between the number of 2-
methylbutyl groups of a single chirality present as terminal
groups and the stereoelectivity has been observed(30). (ii)
In the polymerization of monomers having different optical
purity, using catalysts prepared from aluminum alkyls, stereoelectivity is very small if any, and independent of the optical
purity of the monomer while stereoselectivity exists as shown
by the nonlinear increase of optical rotation with optical
purity(24). (iii) In the copolymerization of (R)(S)-3,7-DMO
with (S)-3-MP with catalysts prepared from tris-isobutyl
aluminum, a remarkable stereoselectivity exists, while stereoelectivity is very small. Therefore, according to the experimental results, we must conclude that stereoselectivity is
substantially originated by the existence of chiral catalytic
centers.

At present it is not possible to say if the chirality of
the catalytic centers is derived from a chiral arrangement of
the substituent around a metal atom (asymmetric transition(39)
or non-transition(40) metal atom) or from the presence of a
chiral group bound to the metal itself(42). The very high
stereoselectivity (up to 90% or more) with respect to the low
stereoelectivity seems to substantiate the former hypothesis.
In fact, it is reasonable to admit that the highly stereoselective catalyst becomes able to induce a rather low
stereoelectivity when dissymmetrically perturbed by a chiral
moiety σ-bound or Π-complexed to a metal atom of the catalytic
complex.

5) *SOME REMARKS ON THE MECHANISM OF STEREOSPECIFIC AND
   STEREOSELECTIVE POLYMERIZATION.*

The existence of chiral active centers in heterogeneous
Ziegler-Natta catalyst, already postulated in 1958(41), has
been experimentally confirmed by the above described experiments involving chiral α-olefins and subsequently by detailed
NMR investigation of the isotactic polymers from propylene(43

SCHEME 9.

Copolymerization of ethylene with a chiral α-olefin (21).

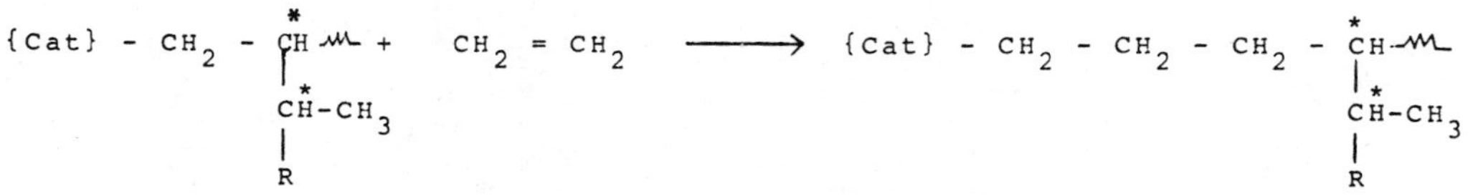

$$\{Cat\} - CH_2 - \overset{*}{C}H \sim\!\!\sim + \quad CH_2 = CH_2 \quad \longrightarrow \quad \{Cat\} - CH_2 - CH_2 - CH_2 - \overset{*}{C}H \sim\!\!\sim$$

with substituents $\underset{R}{\overset{|}{C}H-CH_3}$ (marked $*$) on each.

SCHEME 10.

Two steps mechanism for anionic coordinated polymerization of ethylene (27).

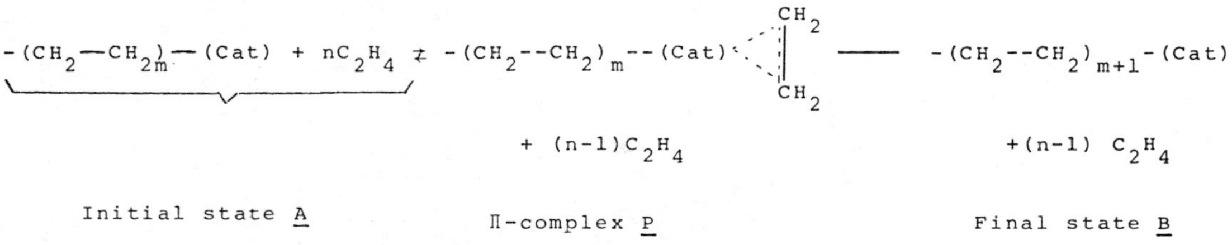

$$-(CH_2\!-\!CH_2)_{\overline{m}}(Cat) + nC_2H_4 \; \rightleftarrows \; -(CH_2\!-\!-CH_2)_m\!-\!-(Cat) \overset{CH_2}{\underset{CH_2}{\cdots}} \quad\longrightarrow\quad -(CH_2\!-\!-CH_2)_{m+1}\!-(Cat)$$

$$+ (n-1)C_2H_4 \qquad\qquad\qquad\qquad\qquad + (n-1)\ C_2H_4$$

Initial state $\underline{A}$ Π-complex $\underline{P}$ Final state $\underline{B}$

and its copolymers with ethylene(44). Taking into account
the chirality of the active centers, the addition of each
single unit to the growing chain can be treated as a case of
asymmetric catalysis by transition metal complexes, a field
which has been intensively investigated in the last five
years.

   In this section an attempt is made to rationalize the
results obtained in the stereoselective polymerization based
on the same approach successfully used for the interpretation
of the asymmetric hydroformylation of α-olefins(45). The
mechanism of both reactions is believed to be based on
coordination of the olefin to a transition metal atom followed
by insertion into a metal-carbon or metal-hydrogen bond as
supported by several experimental findings(27,46,47). The
above analogy is strictly valid for soluble Ziegler-Natta
catalysts and can tentatively be extended to heterogeneous
catalysts, due to parallelism which exists for the basic
steps of the mechanism of homogeneous and heterogeneous
catalysis(48).

5.1) A Possible Reaction Path for a Single Step of Ethylene and Propylene Polymerization

   Before discussing a possible mechanism for the stereospecific and stereoselective polymerization of racemic α-
olefins, it is convenient to examine the much simpler case of
ethylene and propylene. According to the generally accepted
(27) mechanism*, the addition of each single monomer molecule
to the growing chain in the polymerization of ethylene by
Ziegler-Natta catalyst can be suitably described by two
successive reaction steps. The first is the formation of a
Π-complex of ethylene with a transition metal atom of the
catalytic center (Scheme 10a), and the second is the insertion
reaction of the complexed olefin into the bond between a metal
atom of the catalytic center and a methylene group of the
growing chain (Scheme 10). Very little is known about the
activation energy of the two steps (Fig. 3). By analogy,

---

*
Another possible reaction mechanism which is not discussed
here is the following:

$$\sim\!\!\sim(CH_2\text{-}CH_2)_m - Cat \overset{CH_2}{\underset{CH_2}{\langle\ \vdots\ }} \xrightarrow{+C_2H_4} \sim\!\!\sim(CH_2\text{-}CH_2)_{m+1} - Cat \overset{CH_2}{\underset{CH_2}{\langle\ \vdots\ }}$$

in which the olefin insertion and the complexation of the next
olefin molecule occur simultaneously.

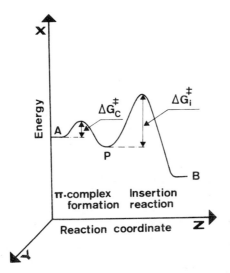

Fig. 3. Reaction path for the two steps addition of a single ethylene unit to a growing chain (see Scheme 10); $\Delta G_i^{\ddagger}$ arbitrarily assumed higher than $\Delta G_c^{\ddagger}$. (For the meaning of A, P and B see Scheme 10).

with other addition reactions of olefins catalyzed by transition metal complexes, it seems reasonable to assume that the Π—complex formation occurs more rapidly than the insertion reaction, and that the complexation equilibrium is promptly achieved under reaction conditions(41,47,49).

The case for α-olefins' polymerization is more complicated. In fact, due to their prochiral nature, two π-complexes (P_1 and P_2 for propylene in Scheme 11) are formed depending on the face, *re-re* or *si-si*, (50) attached to the metal atom in the catalytic complex. Moreover, the insertion reaction can be either completely stereospecific or, because of the non-synchronous addition of the metal atom of the catalytic center and of the growing chain end to the double bond, only partially stereospecific.

Let us assume first that the former case occurs and that a *cis* addition takes place(51). If the catalytic centers are chiral ([Cat]*), the two possible π-complexes at a single center (P_1 and P_2 are diastereomeric),the difference of free energy of formation being ΔG°. This type of diastereomerism in olefin π-complexes has been demonstrated in *cis*(52,53) and *trans*(53) (olefin)(amine)-dichloro-platinum(II) complexes. When, in the *cis* complexes, (S)-α-phenylethylamine is used as ligand, a prevalence up to 20% of one diastereomer has been

61

SCHEME 11.

Stereochemistry of the insertion in a growing chain of a single propylene unit.

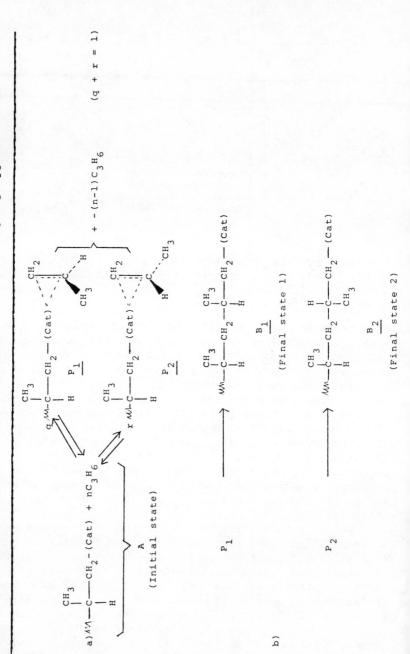

62

reached at the equilibrium(53), although the Pt-atom is not a chiral center. If a rapid complex formation and dissociation takes place, the equilibrium is rapidly reached between P_1 and P_2 and a possible reaction path is represented in Fig. 4a. If z is the reaction coordinate, a projection of the reaction path on the xy plane (Fig. 4b) shows that in order to obtain a stereoregular polymer containing 98% (54) of isotactic (meso) diads, the difference of free energy of the two transition states, $\Delta\Delta G^{\neq}$ must be of the order of magnitude of 2500 cal/mol at 25°C. This difference arises both from the free energy difference ΔG° between the two π-complexes and from the two activation energies $\Delta G^{\neq}_{i1}$ and $\Delta G^{\neq}_{i2}$ of the insertion reaction starting with P_1 and P_2, respectively.

$$\Delta\Delta G^{\neq} = \Delta G^{\neq}_{i2} - \Delta G^{\neq}_{i1} + \Delta G^{\circ} = \Delta\Delta G^{\neq}_{1,2} + \Delta G^{\circ} \qquad [3]$$

The limiting cases can be considered: (a) $\Delta\Delta G^{\neq}_{1,2}$ is zero; in this case $\Delta\Delta G^{\neq}$ is equal to ΔG° and the stereospecificity is determined only by the different thermodynamic stability of the two π-complexes P_1 and P_2; (b) $\Delta\Delta G^{\neq}_{1,2}$ is different from 0; in this case if ΔG° is negligible with respect to $\Delta G^{\neq}_{1,2}$ the stereospecificity is determined by the difference of activation energy of the addition of the metal atom and growing chain end to the *re-re* face (P_1) or to the *si-si* face (P_2) of the olefin. If both $\Delta\Delta G^{\neq}_{1,2}$ and ΔG° are not negligible they can affect stereospecificity either in the same or in the opposite direction.

No data are available to establish the extent to which the stereoregularity is controlled by ΔG° or by $\Delta\Delta G^{\neq}_{1,2}$. In the asymmetric hydroformylation the results obtained with aliphatic α-olefins can be explained by assuming $\Delta G^{\neq} = \Delta G^{\circ}$. However, in this case the process takes place in solution and $\Delta G^{\neq}$ with the catalysts used up to now, is smaller than in the polymerization, as shown by the relatively low asymmetric induction achieved. Data on the nature and stability of the π-complexes involved in the polymerization as well as on the activation energy of the insertion reactions in heterogeneous phase are necessary for a quantitative insight into the factors regulating stereospecificity in anionic coordinated polymerization.

If the insertion reaction is not stereospecific, the addition of catalyst and growing chain end to the complexed olefin being for instance nonsynchronous, formation of atactic macromolecules can take place even if the chiral catalytic site is able to distinguish between the two prochiral faces of the olefin.

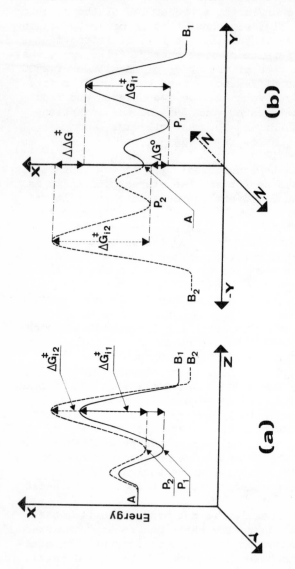

Fig. 4. Reaction path of the two steps addition of a single propylene unit to a growing chain with completely stereospecific insertion step: (a) (———) formation of an isotactic (meso) diad; (– – –) formation of a syndiotactic (racemic) diad (projection on the xz plane). (b) projection of the above reactions paths on the x,y plane. B_1, isotactic (meso) diad, has been arbitrarily assumed to be less stable than B_2, syndiotactic (racemic) diad. For the meaning of A, B_1, B_2 P_1 P_2 see Scheme 11.

5.2) <u>Stereoselective Polymerization of Racemic α-Olefins.</u>

In the case of racemic α-olefins containing one asymmetric carbon-atom, four diastereomeric π-complexes are formed by complexation to a metal atom of a single chiral center (Scheme 12, P_{1R}, P_{2S}, P_{1S}, and P_{2R}). Let us consider a limiting case suggested by the preceding discussion on propylene polymerization and by the results of the asymmetric hydroformylation(45). In this particular case the formation of a completely isotactic macromolecule is considered in which the free energy of activation for the insertion reaction is practically independent of the complexed face of the olefin, that is, $\Delta\Delta G^{\neq}_{1,2}$ in eq. 3 is zero. In this case the insertion rate of the olefin in the growing chain is the same or very similar starting with P_{1R} or P_{1S} where the *re-re* face is attached to the metal, or correspondingly with P_{2R} or P_{2S} where the *si-si* face is complexed. Therefore, assuming that $\Delta\Delta G^{\neq} = \Delta G°$, the isotacticity will be due, as in the previous case of propylene, to the large prevalence of one of the above π-complex couples ($\Delta G° \approx 2500$ cal/mol at 25°). If for instance, the *re-re* face is prevailingly bound to the catalyst (50), two diads B_{RR} and B_{RS} can be obtained by addition of an (R)- or (S)-olefin molecule, respectively, to a growing chain where the last inserted unit is (R) (Scheme 13). The reaction paths are represented in Fig. 5, where it has arbitrarily been assumed that the isotactic (meso) diad formed from the same antipode B_{RR} is less stable than the B_{RS} formed by the two different antipodes. In this simple case, if the free energy of activation for the insertion of the complexed antipodes is the same ($\Delta G^{\neq}_{11(S)} = \Delta G^{\neq}_{11(R)}$), then the stereoselectivity depends on the value of the equilibrium constant K_{re} between P_{1R} and P_{1S} in which the *re-re* face of the (R)- or (S)-monomer is complexed to the metal, respectively (Fig. 5). The free energy difference between these two complexes, $\Delta G°_{re} = -RT\ln K_{re}$, can be divided into two parts, one being determined by the interactions of the chiral carbon atom of the olefin with the chiral active site, and the other being associated only with the intrinsic structure of the chiral olefin and is independent of the chiral nature of the catalyst. The latter possibility has been investigated for the epimerization of the type $P_{1R} \rightleftarrows P_{2R}$ of square planar complexes *cis* (amine) (olefin)-dichloroplatinum(II) with a nonchiral amine and an optically active α-olefin. It was found(56) that the prevalence of one diastereoisomer at the equilibrium is of the same order of magnitude as the average prevalence of units from one single antipode in the macromolecules obtained by stereoselective polymerization of the corresponding racemic

SCHEME 12.

Diastereomeric π-complexes formed by complexation of the two prochiral faces of an (R)- or (S)-α-olefin to a chiral catalytic center. $R_{(R)}$ and $R_{(S)}$ indicate alkyl groups with an (R)- or (S)-asymmetric carbon atom, respectively).

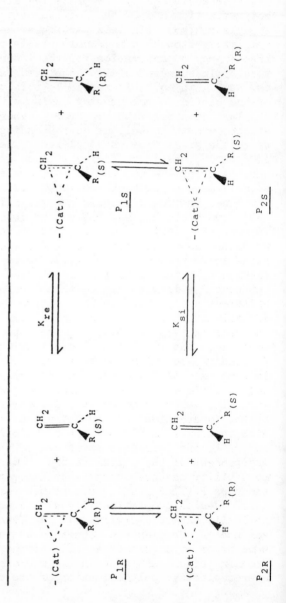

SCHEME 13.

Two steps mechanism for isotactic polymerization of a racemic α-olefin on a catalytic chiral center bearing a growing chain where the ultimate monomeric unit is (R).

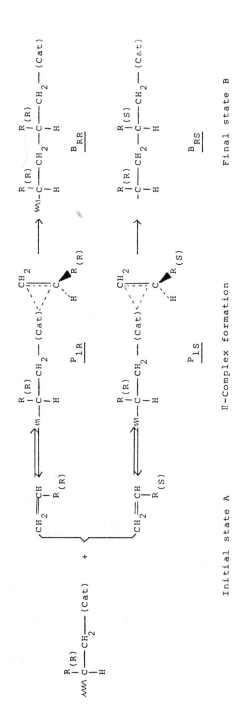

67

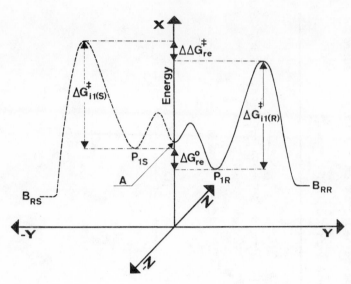

Fig. 5. Projection on the x,y plane of the reaction path for ideal isotactic polymerization of a single (R)- and (S)-α-olefin molecule on a chiral center preferring the (R)-monomer; B_{RR} arbitrarily assumed less stable than B_{RS}. (For the meaning of the symbols see Schemes 12 and 13).

α-olefins. This indicates that direct interactions between chiral centers of monomer and catalyst are not very important and can be ignored. In this case, if a chiral center chooses the *re-re* face, to build up an isotactic macromolecule, at this center the antipode will be preferentially polymerized in which the *re-re* face prevails at the epimerization equilibrium; the extent of preference for one antipode is determined by $\Delta\Delta G^{o\ddagger}_{re}$ which, according to our oversimplified treatment, is equal or very near to ΔG^o_{re}. If the assumptions made are substantially correct, it is possible to conclude that the stereospecific polymerization of racemic α-olefins to isotactic polymers must be stereoselective(57) when the absolute values of $\ln K_{re}$ (and $\ln K_{si}$) (Scheme 13) are sufficiently high.

In conclusion, the stereoselectivity in the isotactic polymerization can be explained on the basis of the difference in thermodynamic stability of the complexes involving the same prochiral face of the two antipodes of a racemic α-olefin. The existence of a certain degree of stereoselectivity in the nonstereospecific polymerization of racemic α-olefins can thus be explained by the presence of catalytic centers able to distinguish first between the two prochiral faces of the olefin and then between the two antipodes, but giving rise to a nonstereospecific insertion reaction.

The parallel decrease of stereoselectivity and stereospecificity shows that centers giving a nonstereospecific
insertion reaction have also a smaller capacity for distinguishing between the monomeric antipodes. The most evident
example is given by the soluble Ziegler–Natta catalysts prepared from tetrabenzyl-titanium and tribenzylaluminum which
give atactic polymers from (R)(S)–4–MH with low stereoselectivity if any(18).

6) FINAL REMARKS

The main results arising from the investigation of the
polymerization of α-olefins can be summarized as follows:

(1) Insoluble Ziegler–Natta catalysts contain chiral
centers which are responsible for the stereospecificity and
for the stereoselectivity of the polymerization of racemic
α-olefins.

(2) If an isotactic stereospecificity arises essentially
from the preferential complexation of one prochiral face of
the α-olefin, followed by a stereospecific insertion process,
the stereospecific polymerization of racemic α-olefins leading
to isotactic polymers can be expected to be stereoselective,
at least when the asymmetric carbon atom is in α or β position
with respect to the double bond and when the interaction
between the chirality center of the olefin and the chiral
catalytic site is negligible.

(3) A non-completely stereospecific course of the insertion reaction might be responsible for the existence of
stereoselectivity in polymerization processes occurring with
low stereospecificity.

(4) The low stereoelectivity observed when the stereospecific catalysts prepared from an optically active metal
alkyls are used, indicates that the small difference in polymerization rate of the two antipodes arises from a modification of the environment of the enantiomeric catalytic centers
rather than from a prevalence of catalytic centers having the
same type of chirality.

ACKNOWLEDGMENT

The authors wish to express their thanks to Prof. P.
Salvadori, Dr. G. Consiglio and Dr. R. Lazzaroni for helpful
discussions. F.C., C.C. and E.C. thank C.N.R. (Roma) for the
financial support.

LITERATURE CITED

(1) E.L. Eliel, "Stereochemistry of Carbon Compounds" McGraw-Hill Book Co., Inc., and Kogakusha Co., Ltd., 1962, p. 434.

(2) P. Pino, F. Ciardelli, G.P. Lorenzi, and G. Natta, *J. Amer. Chem. Soc. 84,* 1487 (1962).

(3) P. Pino, F. Ciardelli, and G. Montagnoli, *J. Polymer Sci. C, 16,* 3265 (1968).

(4) G. Natta, P. Pino, G. Mazzanti, P. Corradini, and U. Giannini, *Rend. Acc. Naz. Lincei (VIII), 19,* 397 (1955).

(5) P.L. Luisi, G. Montagnoli and M. Zandomeneghi, *Gazz. Chim. Ital., 97,* 222 (1962).

(6) P. Pino, G. Montagnoli, F. Ciardelli, and E. Benedetti, *Makromol. Chem., 93,* 158 (1966).

(7) T. Tsuruta, *J. Polymer Sci. D,* 179 (1972).

(8) E. Chiellini, G. Montagnoli, and P. Pino, *J. Polymer Sci. B, 7,* 121 (1969).

(9) L. Porri, and D. Pini, *Chim. Ind.* (Milan).

(10) G. Natta, *J. Inorg. Nuclear Chem., 8,* 589 (1958).

(11) J.D. Morrison, H.S. Mosher "Asymmetric Organic Reactions", Prentice Hall Inc., Englewood Cliffs, N.J., pg. 30 (1971).

(12) S. Inoue, T. Tsuruta, J. Furukawa, *Makromol. Chem., 53,* 215 (1962).

(13) P. Pino, F. Ciardelli, G. Montagnoli, and O. Pieroni, *J. Polymer Sci. B, 5,* 307 (1967).

(14) F. Ciardelli, E. Benedetti, and O. Pieroni, *Makromol. Chem., 103,* 1 (1967).

(15) G. Montagnoli, D. Pini, A. Lucherini, F. Ciardelli, and P. Pino, *Macromolecules, 2,* 684 (1969).

(16) P. Neuenschwander, Dissertation E.T.H.-Zürich (1973).

(17) E. Chiellini and M. Marchetti, *Makromol. Chem., 169,* 59 (1973).

(18) A. Oschwald, Dissertation E.T.H.-Zürich (1974).

(19) O. Pieroni, G. Stigliani, and F. Ciardelli, *Chim. Ind.* (Milan)*52,* 289 (1970).

(20) F. Ciardelli, A. Zambelli, P. Locatelli, and M. Marchetti, IUPAC Symposium on Macromolecules, Helsinki-1972, Prep. Vol. 2, p. 901.

(21) F. Ciardelli, P. Locatelli, M. Marchetti, and A. Zambelli, *Makromol. Chem., 175,* 923 (1974).

(22) P. Pino, F. Ciardelli, and G.P. Lorenzi, *J. Amer. Chem. Soc., 85,* 3888 (1963).

(23) F. Ciardelli, E. Benedetti, G. Montagnoli, L. Lucarini, and P. Pino, *Chem. Comm., 285* (1965).

(24) F. Ciardelli, G. Montagnoli, D. Pini, O. Pieroni, C. Carlini and E. Benedetti, *Makromol. Chem., 147,* 53 (1971).

(25) G. Natta, L. Porri, and S. Valenti, *Makromol. Chem.*, *67*, 225 (1963).

(26) P. Pino, F. Ciardelli, and G.P. Lorenzi, *J. Polymer Sci. C*, *4*, 21 (1963).

(27) G. Natta, P. Pino, E. Mantica, F. Danusso, G. Mazzanti, and M. Peraldo, *Chim. Ind.* (Milan), *38*, 124 (1956).

(28) P. Pino, F. Ciardelli, and G.P. Lorenzi, *Makromol. Chem.*, *70*, 182 (1964).

(29) C. Carlini, H. Bano, and E. Chiellini, *J. Polymer Sci. A-1*, *10*, 2803 (1972).

(30) F. Ciardelli, C. Carlini, G. Montagnoli, L. Lardicci, and P. Pino, *Chim. Ind.* (Milan), *50*, 860 (1968).

(31) H. Ringger, P. Pino, unpublished results.

(32) S. Croce, Thesis, University of Pisa, Italy, 1974.

(33) A. Faucher, and F.P. Reding in "Crystalline Olefins Polymers", R.A.V. Raff and K.W. Doak, Ed., Intersc. Publ., 1965, pp. 695-100.

(34) F. Ciardelli, C. Carlini, and G. Montagnoli, *Macromolecules*, *2*, 296 (1969).

(35) P. Pino, F. Ciardelli, G.P. Lorenzi, and G. Montagnoli, *Makromol. Chem.*, *71*, 207 (1973).

(36) E. Chiellini, *Macromolecules*, *3*, 527 (1970).

(37) C. Carlini, Thesis, University of Pisa, July 1965.

(38) O. Bonsignori, P.L. Luisi, and U. Suter, *Makromol. Chem.*, *149*, 29 (1971).

(39) H. Brunner, *Angew. Chem.*, *83*, 274 (1971); *Int. Ed.*, *10*, 249 (1971).

(40) C. Eaborn, and I.D. Varma, *J. Organometallic Chem.*, *9*, 377 (1967).

(41) P. Cossee, *J. Catalysis*, *3*, 80 (1964).

(42) J.D. Morrison and H.S. Mosher, "Asymmetric Organic Reactions" Prentice-Hall Inc., N.J., Englewood Cliffs, p. 288-291 (1971) and references cited therein.

(43) A. Zambelli, "NMR Basic Principles and Progress", 4, Springer Verlag, Heidelberg, 1971, p. 101.

(44) A. Zambelli, G. Gatti, C. Sacchi, W.O. Crain, Jr., and J.D. Roberts, *Macromolecules*, *4*, 475 (1971).

(45) P. Pino, G. Consiglio, C. Botteghi, C. Salomon, *Adv. Chem. Series*, *132*, 295 (1974).

(46) R.F. Heck, and D.S. Breslow, *J. Amer. Chem. Soc.*, *83*, 4023 (1961).

(47) J. Boor, Jr., *Ind. Eng. Chem. Prod. Res. Dev.*, *9*, *(4)*, 437 (1970).

(48) H. Heinemann, *Chem. Tech.*, 286 (1971).

(49) R. Cramer, *J. Amer. Chem. Soc.*, *87*, 4717 (1965).

(50) K.R. Hanson, *J. Amer. Chem. Soc.*, *88*, 2731 (1966).

(51) A. Zambelli, A.L. Segre, M. Farina, and G. Natta, *Makromol. Chem.*, *110*, 1 (1967).

(52) G. Paiaro, P. Corradini, R. Palumbo and A. Panunzi, *Makromol. Chem.*, *71*, 184 (1964).

(53) P. Pino, R. Lazzaroni and P. Salvadori, in preparation.

(54) F.C. Stehling, *J. Polymer Sci. A2*, 1815 (1964).

(55) P. Pino, R. Lazzaroni and P. Salvadori, *Inorg. Chim. Acta*, III International Symposium Venice 1970, Proc. C2.

(56) P. Pino, R. Lazzaroni, and P. Salvadori, *Chimia*, *24*, 152 (1970).

(57) R. Lazzaroni, P. Salvadori, and P. Pino, *Chem. Comm.*, 1164 (1970).

The Number of Active Sites for the Polymerization of Ethylene, Propylene and Butene-1 by Ziegler-Natta Catalyst

H. SCHNECKO, K.A. JUNG and W. KERN

Department of Chemistry
University of Mainz
D-65 Mainz, FRG

SUMMARY

In the heterogeneous catalyst system $TiCl_3/Al(C_2H_5)_2Cl$ the number of active sites, C^*, is determined by quenching the polymerization of ethylene, propylene and butene-1 with tritiated butanol as a function of time. Peculiarities of the quenching reaction are pointed out. Together with rate and viscosity measurements kinetic constants have been calculated. It can be shown that the catalyst system is very sensitive towards the individual monomer at the beginning as well as during the polymerization.

1) INTRODUCTION

In spite of a large number of publications, many questions concerning the mechanism of heterogeneous Ziegler-Natta polymerizations remain unsolved. The most thorough recent investigation has been conducted by Burfield, Tait and co-workers[1] on the catalyst system $VCl_3/Al-(alkyl)_3$ with 4-methyl pentene-1 as monomer; they attempted to give a general kinetic scheme and included active center determination. Partly due to the problems of determination, this entity – concentration of active sites, C^* – has been controversial for quite some time[2]. Recent results[3-5] confirm the broad spectrum of figures obtainable. In our view this is a consequence of the heterogeneous nature of the catalyst surface which varies with the catalyst preparation and other

73

experimental conditions.

It is the purpose of this paper to show that even under identical initial conditions, C* of the same catalyst is different for monomers in the homologue series: ethylene, propylene and butene-1. Technique used here to determine C* is the quenching of a polymerization by tritiated alcohol, as introduced some years ago into the field of Ziegler-Natta polymerizations by Feldman and Perry(6). Previously, we have shown that labelling by iodine does not give quantitative conversion in our system in spite of the accuracy attainable by either microanalytical titration or neutron activation analysis(7).

2) *EXPERIMENTAL*

Details of the experimental set-up and procedures have been described previously(8). The system consists of a closed apparatus which allows the measurement of the uptake of gaseous monomers at pressures <1 atm - in this case ethylene and propylene at 700 torr, and butene-1 at 500 torr - in 100 ml of Sinarol (high boiling saturated aliphatic hydrocarbons ex Shell). The polymerization conditions used in the present work are the following: temp. is 60°C, time is 120 (or 60) min, and the catalyst system is $TiCl_3$ (Stauffer AA)/$AlEt_2Cl$ ($Et=C_2H_5$). The onset of the polymerization was determined from the saturation time of the monomers as well as from extrapolation of time-conversion curves. They are in agreement to 1 min. for ethylene and 4 min. for both propylene and butene-1. The polymerization was quenched by tritiated n-butanol (specific activity 543/µ Cie/mmol.). The samples were combusted in O_2/methanol after Schöniger and the tritiated H_2O formed was determined by liquid scintillation counting (Tri-Carb 3380). Details of this procedure as well as of the viscosity measurements were previously described(7,9).

3) *BASIC CONSIDERATIONS*

As outlined previously(2), end-group determination in anionic Ziegler-Natta polymerizations can principally be achieved in 2 ways: a) from the initiation b) from the termination reaction; the latter process was used here. Both methods have their merits. However, both are susceptible to chain transfer, e.g., with the Al-compound, thus increasing the labelled polymer molecules(2).

In the case of radioactive tagging via the initiator, i.e. the alkyl group (by ^{14}C) of the Al-compound, there is an additional complication if the transition metal is reduced by the Al-alkyl(10); radioactive olefin can then be released and copolymerized, simulating too high a C^*-concentration(11).

It is therefore imperative to separate inactive, non-growing metal-polymer-bonds (MPB', resulting from transfer with Al-alkyl) from growing polymer chains connected to an active site (C^*); this is usually done by determination of the total number of metal-polymer bonds (MPB), which is the sum of inactive and active chains,

$$MPB = C^* + MPB' \qquad [1]$$

as a function of time. Extrapolation to the beginning of the polymerization renders C^* because then no transfer has occurred yet. In addition, Burfield and Tait(1) have determined the dependence of MPB with catalyst and monomer concentration and used the results to determine transfer rates.

It should be noted that to our knowledge there have been only 2 attempts where simultaneous tagging of polymerizing chains has been attempted via initiation *and* termination, both with soluble initiator systems. Some years ago, Chien (11) used $Al(^{14}CH_3)_2Cl/(C_5H_5)_2TiCl_2$ initiation and $^{131}I_2$ quenching in ethylene polymerization. He found first order initiation and second order termination, but did not attempt to separate MPB' and C^*. Incidently, in butene-1 polymerization with our heterogeneous system, a mixed order for termination was reported(12), indicating active sites of variable stability. Recently, Keim, Christman, Kangas and Keahey(13) utilized $Al_2(^{14}C_2H_5)_3Cl_3/VOCl_3$ initiation and i-$C_3H_7O^3H$ termination for ethylene-propylene copolymerizations, but unfortunately the conditions were such that it was difficult to obtain high accuracy and meaningful agreement in the results.

4) *KINETIC EQUATIONS*

For the highly active system used here we were unable to work out a detailed kinetic treatment that was valid for all 3 monomers. Instead the basic equations developed by Natta and Pasquon(14) for propylene were adopted(2,15). The rate constant k_p for chain growth was obtained from

$$R_p = k_p \cdot [M][C^*] \qquad [2]$$

with R_p = rate of polymerization and $[M]$ = concentration of monomer. R_p was evaluated from conversion-time-curves normalized for catalyst concentration,(Fig. 1).

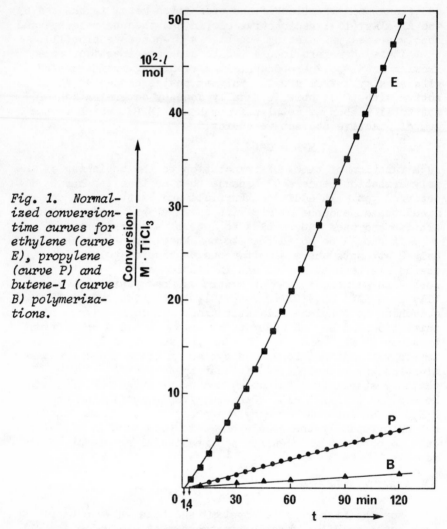

Fig. 1. Normalized conversion-time curves for ethylene (curve E), propylene (curve P) and butene-1 (curve B) polymerizations.

Count(16) has shown that the overall chain transfer constant K_{tr} is obtained from the total number of MPB as a function of polymer formed, provided the number of transferred polymer molecules MPB' is proportional to conversion Q [mol/l] of monomer. Thus, if

$$MPB' = K_{tr} \cdot Q \qquad [3]$$

is introduced into eq. [1], then

$$MPB = C^* + K_{tr} \cdot Q \qquad [1a]$$

76

TABLE 1.

Tritium isotope effect in the termination of ethylene, propylene and butene-1 polymerizations as measured by n-BuOH/BuO³H.

Nr.	Monomer	$\dfrac{BuO^3H}{Me-X^a}$	Radio-activity[b] (μCie/g)	Apparent isotope effect	MPB $\cdot$ 10^3 (mmole/mmole TiCl$_3$)
1	ethylene	1.3:1	4.29	2.60±0.21	87.6
2	propylene	1:1	5.81	1.63±0.05	61.6
3	propylene	1.3:1	3.95	2.36±0.21	58.2
4	butene-1	1:1	5.93	1.37	43.7

[a] X = R, Cl; total number of hydrolysable metal bonds in the system. [b] Polymerization time: 60 min.

77

Furthermore,

$$Q = \int_{o}^{t} R_p \, dt \qquad [4]$$

and from eq. [3] and [4] one derives

$$\frac{dMPB'}{dt} = R_{tr} = R_p \cdot K_{tr} \qquad [5]$$

(with MPB' = metal polymer bonds formed by transfer). With some approximations(14) the individual rate constant k_{tr} for the transfer reaction with the cocatalyst $AlEt_2Cl$ can be estimated from R_{tr}:

$$R_{tr} = k_{tr} \cdot [C^*][Al]^{1/2} \qquad [6]$$

One might argue that in this expression a number of factors have not been taken into account; however, it could be shown by iodine addition to the terminated polymer that at least the contribution of transfer reactions with a monomer are on the order of $\lesssim 10\%$ (7).

Finally, the determination of the lifetime L of the growing macromolecules for stationary conditions is possible with

$$L = \frac{[C^*] \cdot \overline{P}_{n\infty}}{R_p} \qquad [7]$$

provided the limiting value of the average degree of polymerization $\overline{P}_{n\infty}$ for infinite polymerization time can be estimated(9,14).

5) *RESULTS*

5.1) Tritium quenching

The crucial nature of the isotope effect (k_H/k_{3_H}) for the quenching reaction with tritiated alcohol has been pointed out previously(2). Here, it suffices to say that differences in the value of the isotope effect found by various researchers may again be due to the complexity of the heterogeneous system and the different reactivities of the various metal-organic species. In order to stress our claim that it is necessary to determine the isotope effect under the specific conditions for a particular monomer even with the same catalyst system, Table 1 gives experimental data for the 3 monomers we are concerned with. It is seen that the "apparent" isotope effect varies for ethylene, propylene and butene-1. Runs No. 2 and 3 show that the apparent isotope effect in propylene polymerization is changed by about 50% with 30% change in the tritiated butanol concentration.

That this dependence is real is substantiated by the agreement in overall MPB (61.6 and 58.2) shown in the last column for Nrs. 2 and 3.

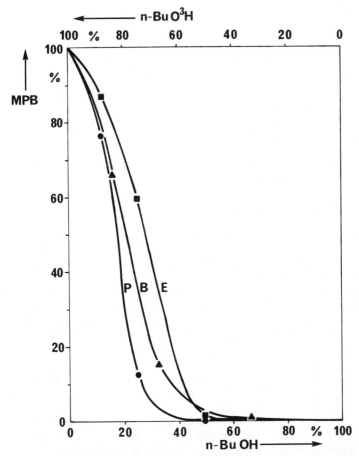

Fig. 2. *Relative number of metal polymer bonds (MPB, in %) as a function of the fraction of tritiated n-butanol (n-BuO³H) added in the quenching of ethylene (curve E), propylene (curve P) and butene-1 (curve B) polymerizations. Completion of the termination by the respective vol-% of n-BuOH after 10 min.*

The above mentioned rate differences in the quenching reaction are illustrated by Fig. 2. Here, the amount of alcohol required to terminate all metal-organic polymer

chains is determined. This is done by first calculating the
amount of alcohol required to destroy all hydrolizable bonds
based on the initial catalyst concentration. Then a fraction
of this amount is added as tritiated butanol and, after 10
mins., complete quenching is accomplished by the addition of
the remaining quantity of inactive butanol. Fig. 2 shows
that in each case more than 97% of the MPB are destroyed with
50% of the calculated amount of the terminating agent. The
sequence of the curves for polyethylene and polypropylene
(curves E and P) is in accordance with their solubility
behaviour (cf. section 6). The position of curve B, for poly-
butene-1 is somewhat unexpected because the much higher solu-
bilization of this polymer should facilitate the access of
the quenching reagent and shift the curve to the left of the
polypropylene curve. In any case, it must be concluded from
the experiments that polymeric chains can react faster than
the other metal bonds.

5.2) Metal polymer bonds (MPB)

From Fig. 1 the large differences in the polymerization
rates of the 3 monomers are apparent, i.e. the much faster
rate of ethylene as against its 2 higher homologue olefins,
propylene and butene-1. In addition, it can be seen from the
figure that the rate of the ethylene polymerization increases
during the first 90 min, whereas the polymerization of propylene essentially proceeds at a constant rate; depending on
catalyst activity, the rate curves for the latter show a more
or less pronounced maximum at the beginning of the polymeri-
zation(17). Due to the ordinate scale of Fig. 1 this is,
however, not clearly discernible. The same is true for
butene-1 polymerizations which show a rate decrease in the
initial period.

The results of the quenching experiments are found in
Fig. 3 where the MPB are plotted vs. amount of polymer
formed; very similar curves are obtained if time is taken as
the abscissa. In our system with practically unlimited monomer supply it is hardly possible to plot against increments
of % conversion as has been done by Burfield and Tait(1). In
any case, the different curvatures in Fig. 3 cannot be
expected to obey a common relationship. The intercepts of
the 3 curves do, however, give C*, the concentration of active
centers at the beginning of the polymerization. This extrapolation is quite accurate because the initial part of each
curve is fairly linear. The values for C* thus obtained are
given in Table 2 (section 5.3).

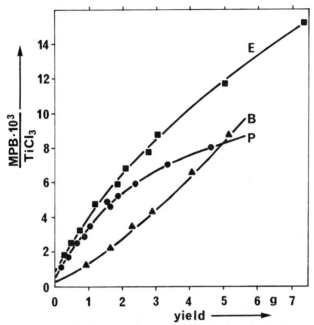

*Fig. 3. Normalized metal-polymer bond con-
centration (MPB) vs. yield of polyethylene in g (curve E), polypropylene
(curve P) and polybutene-1 (curve B).*

The MPB concentrations increase with conversion for all
3 monomers. From Fig. 1 it is clear that this is not due to
increasing C^*-figures, except for ethylene which has a
slightly accelerating rate of polymerization. The increase
in MPB is, in fact, indicative of the important chain transfer
with the cocatalyst as expressed in eq. [6]. It can be
derived from the shape of the curves of Fig. 3 that such
transfer reactions decrease with conversion in the case of
ethylene and propylene, whereas they increase for butene-1.

With the values of MPB, it is possible to calculate the
number-average degrees of polymerization P_n (9); they are
plotted against time for the 3 monomers in Fig. 4. The
increase of the ethylene and propylene curve and the decrease
for butene-1 substantiate the claim made above for chain
transfer. Measurement of intrinsic viscosities were used to
calculate M_w (9). In contrast to M_n (corresponding to P_n),
increasing curves were found in each case. Both values
allowed calculation of the uniformity M_w/M_n which is depicted
in Fig. 5 as a function of polymerization time. It can be

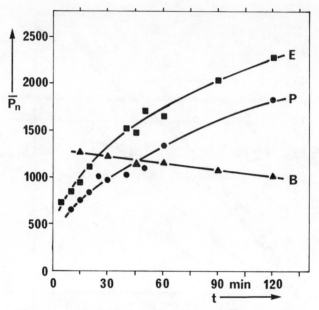

*Fig. 4.   Number average degree $\overline{P}_n$ of polymer-
ization as a function of time in the
polymerization of ethylene (curve E),
propylene (curve P) and butene-1
(curve B).*

seen that the distribution is very broad, in particular for
polyethylene ($M_w/M_n \sim 20$); for polypropylene, uniformity has
been reported to range between 8 and 15(2,18). For both
polymers one observes a slight decrease in the values with
conversion or time respectively. On the other hand, the
polybutene-1 formed initially has a low M_w/M_n which doubles
during the course of the polymerization (Fig. 5, curve B).

5.3) <u>Kinetic data</u>

Finally, Table 2 summarizes some of the important data
calculated from Figs. 1-5 and, in addition, gives kinetic
rate constants calculated from the equations outlined in
section 4 (*KINETIC EQUATIONS*). A common quasi-stationary
state for all 3 monomers which lasts for longer periods of
time is not observed with our system. This might partly be
due to the experimental conditions of a) a highly active
catalyst and b) constant monomer concentration. In view of
this, the calculations were restricted to the initial (10-
30 min.) and the final period (90-120 min.) of the polymeri-

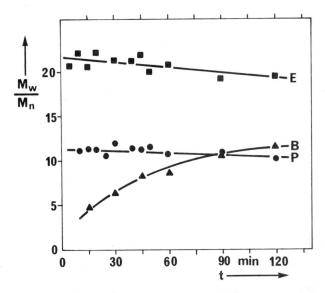

Fig. 5. Polydispersity (M_w/W_n) as a function of time for ethylene (curve E), propylene (curve P) and butene-1 (curve B) polymerizations.

zation. This is, of course, not a rigorous kinetic practice. However, it appears justified in view of the complexity and non-uniformity of this heterogeneous system. Moreover, it is to be preferred to a simple average for the total period. Our analysis at least reveals some changes in the magnitude of the individual rates and constants.

The rate constant for chain propagation, k_p, has been exempted from this procedure since it is difficult to envisage a dependence of k_p with time. This does not exclude the existence of different k_p's in such a system arising from the presence of C*'s with different activity as has been postulated previously(19) and recently confirmed by morphological studies(20). The k_p-value for ethylene is in good agreement with the literature(21) whereas that for propylene is lower than described(22). The large activity differences between the 3 monomers as already shown in Fig. 1 are expressed in the ratio of k_p-values in Table 2 (11 : 2.5 : 1 for ethylene : propylene : butene-1). A similar comparison of the concentration of active sites C* at the start of the polymerization yields a ratio of 2 : 1.5 : 1 for ethylene : propylene : butene-1 in very good agreement with results from previous investigations where activity changes by partial catalyst

TABLE 2.

Polymerization constants for ethylene, propylene and butene-1.

Constant	Pol. time (min)	Ethylene	Propylene	Butene-1
$[TiCl_3] \cdot 10^3$ (mole/l)		7.45	7.45	10.34
$[Al(C_2H_5)_2Cl] \cdot 10^3$ (mole/l)		100	100	100
Al/Ti		13.4	13.4	9.7
[M] (mole/l)		0.07	0.24	0.52
k_p (1/mole/sec)		78.1	18.0	7.3
$[C*] \cdot 10^3$ (mmole/mmole $TiCl_3$)	0 / 90–120	6.9 / 10.2	5.3 / 4.7	3.5 / 2.8
$[C*] \cdot 10^5$ (mole/l)	0 / 90–120	5.1 / 7.6	4.0 / 3.5	3.6 / 2.9
$R_p \cdot 10^4$ (mole/l/sec)	0–30 / 90–120	3.8 / 5.7	2.3 / 2.0	1.3 / 1.1
$R_{tr} \cdot 10^7$ (mole/l/sec)	0–30 / 90–120	3.5 / 2.4	2.6 / 0.64	1.1 / 1.2
$K_{tr} \cdot 10^4$	0–30 / 90–120	9.1 / 4.3	11.3 / 3.2	8.3 / 11.0
$k_{tr} \cdot 10^2$ ($l^{1/2}$ $mole^{1/2} \cdot$ sec)	0–30 / 90–120	1.6 / 0.8	1.5 / 0.4	1.0 / 1.3
$\overline{M}_n \cdot 10^{-3}$	∞	86	103	50
$\overline{P}_n$	∞	3071	2452	892
L (min)	∞	9.2	9.5	3.9

poisoning(19) had been studied. During the process of the polymerization there is an increase in C^* for ethylene, whereas there is a decrease for propylene and even more for butene-1 so that during the last period (90-120 min.) the ratio of C^* is increased to 3.6 : 1.7 : 1. (C^* in the final period, 90 to 120 min., was calculated from eq. [2]).

The changes in C^*-concentration are also reflected in the values of R_p. Accepting the validity of eq. [2] and the constancy of k_p (v.s.), one could still argue that R_p variations with time were due to changes in monomer concentration, [M]. But for our conditions it has been shown that the polymerization is not diffusion-controlled(8,9), so that [M] can be considered constant.

The numbers given for the transfer rates, R_{tr}, show a decrease with time for ethylene, a sharp fall for propylene and a slight increase for butene-1. This corroborates the remarks made in discussing Figs. 3 and 4 (section 5.2). With R_p always being 3 orders of magnitude larger than R_{tr} it is obvious that P_n must increase in the case of ethylene and propylene because of the relatively long lifetimes of growing chains, L. The situation is reversed for butene-1, where decreasing R_p and C^* are overbalanced by increasing R_{tr}, so that a fall in P_n results (Fig. 4). It should be noted that for propylene both R_{tr} and K_{tr} in Table 1 are of the same magnitude as described before(22) with the catalyst system.

Finally, Table 2 lists the average lifetime of growing polymer chains, L. For both polyethylene and polypropylene chains average L-values of approximately 10 min. at 60°C are found which are in good agreement with findings by Count(16), but larger than those of Coover, Guillet, Combs and Joyner (22) (in case of propylene) and smaller than those of Grieveson(21) (ethylene); no literature comparison for butene-1 has been found.

6) DISCUSSION

Although it can be assumed that the starting conditions of our system, i.e. at the moment of monomer introduction, are identical in each case it is obvious from this study that one catalyst system induces quite a different polymerization behaviour of the 3 homologue olefins. This is clearly manifested by the different C^*-values at time 0, but it is also apparent from the different behaviour of ethylene, propylene and butene-1 during the course of the polymerization. It must be concluded from the results that reactions take place which change the number of active sites.

In order to explain the differences in C* it can be assumed that the steric requirements for the insertion reaction are not identical for each monomer. This was also deduced from earlier studies on the modification of catalyst activity after aging(2,19). On the other hand, the polymerization itself has a strong influence on the balance of C*, as is evident from some of the contradictory kinetic features for butene-1 as against the 2 lower homologues. Primarily in case of ethylene, the growing chain is a crystalline, insoluble globule which causes break-up of the catalyst aggregates and even crystal scissions(14), exposing new surface areas that can react to form new C*. This has also been claimed to occur in case of propylene(18), but if so, to a much smaller extent under our conditions as judged from the small increase in C* and from earlier studies on the dependence of the polymerization rate on catalyst particle size (17). Polypropylene chains will be less rigid and more solubilized than those of polyethylene.

Polybutene-1 chains, on the other hand, are highly solubilized(23) and most likely are unable to cleave catalyst surfaces. A quantitative treatment of the rate decrease in this case(12) indicates that the destruction of active sites in the course of the polymerization is due to the inhibiting action of $AlEtCl_2$ during the formation of C* and to specific transfer reactions with $AlEt_2Cl$ at C* (16).

In passing it should be noted that additional support for the existence of C*'s with different activity for the individual monomers can be drawn from polymerization rate changes with temperature: the Arrhenius-plot for the system $AlEt_3/TiCl_3$ between 20 and 70°C consists of 2 distinct linear parts with a break at 50°C for butene-1 (24) whereas it is a straight line for ethylene and propylene(25). This has been explained by the greater sensitivity of butene-1 monomer towards the temperature-dependent equilibrium of active center-formation and active center-destruction(19). The latter reaction in this case might consist of a Ti-reduction below the 3-valent state since the break was not observed with $AlEt_2Cl/TiCl_3$ and butene-1 (24). Interestingly, Burfield and Tait(1) have found the same behaviour with a break-point at 47°C for Al (i-bu)$_3$/VCl$_3$ and 4-methyl pentene-1.

The location of C* on the catalyst surface is still an open question(1). The $TiCl_3$ used here has a surface of 22 m^2/g (17); assuming that the cocatalyst is adsorbed or chemisorbed on the $TiCl_3$-surface, its coverage by the active sites at the start of the polymerization can be calculated: with a cross-sectional area of 40 Å for $AlEt_2Cl$ (26) and the values of C_o^* (C^* at t = 0) for the 3 monomers shown in Table 2, figures of 25-50% are derived for the coverage of the available surface. Depending on the structure of the active site (not to be discussed here) there could be an additional contribution from monomer insertion on a growing polymer chain. At any rate, the active surface is increasing during the course of the polymerization with ethylene, almost constant with propylene, and decreasing with butene-1, according to the changes in C^*. In our case, restriction of active site locations to the lateral faces which comprise only ca. 5% of the total surface(27) is at variance with the above figures, but recent estimations have even gone to a complete coverage of the crystal surface by C^* (1,5,28).

It should be pointed out that on a molar basis our C^*-concentration is still low, on the order of ≈ 1 mole-% of $TiCl_3$. This figure is, of course, essentially dependent on the grain size of the crystals; it should be considered as an average figure(2-5) in a broad spectrum; smaller C^*-concentrations(1) can be related to much lower transition metal halide surface areas.

Qualitatively, correlation of the other polymerization data and rate constants reported in section 5 can be discussed on the basis of the above. For instance, the difference in transfer reactions can be related to the solubility behaviour of the polymers in conjunction with the respective rate changes. Transfer by Al-alkyl is increasing in case of butene-1 because its solubility is higher   than that of ethylene and propylene. For the latter two, diffusion of the bulky metal alkyl through a more or less heavy layer of polymer is required. The increasing (or constant) rate for ethylene (or propylene), on the other hand, indicates that such restrictions are not operative for the less bulky monomers.

Differences in transfer as well as in propagation rate constants at various active site positions on the catalyst surface cause the broad molecular weight distribution. Changes in C^* and relatively long lifetimes are additional factors which can be quoted in the individual case(9,15).

In summary, the investigation has brought out the great sensitivity of such a highly active catalyst towards the individual monomer; even the "classical" Ziegler-Natta system $TiCl_3/AlEt_2Cl$ demonstrates its complex nature at the onset and during the polymerization of homologue olefins. In a number of studies, including this one, some of the results are in qualitative agreement for such heterogeneous catalysts. However, it is doubtful whether a unified quantitative picture can be created which is applicable to more than one monomer.   This is borne out by the recent conclusion of Borisova, Fushman, Visen and Chirkov(29) that even for the well-ploughed homogeneous catalyst system $C_5H_5TiEtCl/AlEtCl_2$ "experimental results in various laboratories are so contradictory that together they make still more obscure the reason for the activity of soluble Ziegler-Natta type catalysts".

LITERATURE CITED

(1) D.R. Burfield, P.J.T. Tait et al., *Polymer 13*, 302 ff (1972).
(2) H. Schnecko and W. Kern, IUPAC Macromolecular Symposium Budapest 1969, p. 365; *Chemiker Z. 94*, 229 (1970).
(3) Y.I. Ermakov and V.A. Zakharov, *Usp. Khim. 41*, 377 (1972).
(4) I. Meizlik and M. Lesna, *Vysok. Soed. B 15*, 7 (1973).
(5) E.W. Duck, D. Grant, A.V. Butcher and G.D. Timms *European Polymer J. 10*, 77 (1974).
(6) C.F. Feldman and E. Perry, *J. Polymer Sci. 46*, 217 (1960).
(7) H. Schnecko, K.A. Jung and L. Grosse, *Makrom. Chem. 148*, 67 (1971).
(8) H. Schnecko, M. Reinmöller, K. Weirauch, W. Lintz and W. Kern, *Makrom. Chem. 69*, 105 (1963).
(9) K.A. Jung, PhD. Thesis, Mainz University, 1970.
(10) H. Sinn and F. Patat, *Angew. Chem. 75*, 805 (1963).
(11) J.C.W. Chien, *J. Amer. Chem. Soc. 81*, 86 (1959).
(12) H. Schnecko, W. Lintz and W. Kern, *J. Polymer Sci. A1 5*, 205 (1967).
(13) G.I. Keim, D.L. Christman, L.R. Kangas and S.K. Heahey, *Macromolecules 5*, 217 (1972).
(14) G. Natta and I. Pasquon, *Advan. Catalysis 11*, 1 (1959).
(15) K.A. Jung and H. Schnecko, *Makrom. Chem. 154*, 227 (1972).
(16) A.D. Count, *J. Polymer Sci. C4*, 49 (1964).
(17) H. Schnecko, W. Dost and W. Kern, *Makrom. Chem. 121*, 159 (1969).
(18) V.W. Buls and T.L. Higgins, *J. Polymer Sci. A1 8*, 1037 (1970).
(19) H. Schnecko, M. Reinmöller, W. Lintz, K. Weirauch and W. Kern, *Makrom. Chem. 84*, 156 (1965).
(20) J. Wristers, *J. Polymer Sci. 11*, 1601 (1973).
(21) B.M. Grieveson, *Makrom. Chem. 84*, 93 (1965).
(22) H.W. Coover, J.E. Guillet, R.L. Combs and F.B. Joyner, *J. Polymer Sci. A1 4*, 2583 (1966).
(23) H. Schnecko, P. Freyberg, M. Reinmöller and W. Kern, *Makrom. Chem. 100*, 66 (1967).
(24) W. Lintz, PhD. Thesis, Mainz University 1965.

(25) H. Schnecko, M. Reinmöller, K. Weirauch, V. Bednjagin and W. Kern, *Makrom. Chem.* 73, 154 (1964).
(26) S. Tanaka and H. Morikawa, *J. Polymer Sci.* A 3, 3147 (1965).
(27) L. Rodriguez, *J. Polymer Sci.* A1 4, 1905 (1966).
(28) Y.V. Kissin, S.M. Mezhikovsky and N.M. Chirkov, *European Polymer J.* 6, 267 (1970).
(29) L.F. Borisova, E.A. Fushman, E.I. Vizen and N.M. Chirkov, *European Polymer J.* 9, 962 (1973).

The Number of Propagation Centers in Solid Catalysts for Olefin Polymerization and Some Aspects of Mechanism of Their Action

YU.I. YERMAKOV and V.A. ZAKHAROV

Institute of Catalysis, Novosibirsk 90, USSR

1) INTRODUCTION

The discovery of the Ziegler-Natta catalysts greatly influenced the development of catalytic research and gave rise to drastic increase of the investigations in the field of catalytic polymerization. At the Institute of Catalysis the study of polymerization catalysis was initiated by Prof. Boreskov shortly after the organization of this Institute in Siberia early in the 1960's. At first, for this study the chromium oxide catalyst for ethylene polymerization was chosen. The choice of this catalyst can be ascribed to the reason, that the oxide polymerization systems seemed to be closer to the traditional objects of catalytic studies than the organometallic compounds being then quite exotic. The difference between the oxide catalysts (the Phillips Petroleum Co. catalysts - CrO_3 on oxides and the Standard Oil Co. catalyst - MoO_3 on Al_2O_3) and those of the Ziegler-Natta type seemed to be drastic at that time. But now it is clear that olefin polymerization by all solid catalysts practically follows the same mechanism - through monomer insertion into the transition metal - carbon bond with monomer precoordination. In essence, these two types of catalysts differ in a manner of the formation of propagation centers - the surface compounds with an active metal - carbon bond.

In the case of oxide catalysts the formation of propagation centers proceeds by the interaction of a solid catalyst with a monomer, subsequent polymerization also proceeds without a second component - organometallic cocatalyst. The systems of this type can be defined as one-component. The traditional Ziegler-Natta catalysts are two-component. The propagation center formation in these catalysts and subsequent polymerization occur in the presence of organometallic co-catalysts.

In the past decade new solid catalysts of both types have been obtained. In the field of one-component catalysts, active samples of titanium dichloride(1,2) and other transition metal halides were found to initiate polymerization without organometallic compounds. A new method was also developed to obtain one-component catalysts by the interaction of transition metal organometallic compounds with oxide supports(3-5). In the field of two-component catalysts the most significant advances have been made by the development of high activity supported ethylene polymerization catalysts containing titanium halides(6).

To understand the mechanism of action of solid polymerization catalysts the precise data on the number of propagation centers and propagation rate constants are of great importance. In this paper the results of the determination of these values obtained at our Laboratory by using radiotracer quenching techniques will be given, and some aspects of the mechanism of action of different catalysts (oxide, supported organometallic and titanium halides) on the basis of these results will be considered.

2) *THE DETERMINATION OF THE NUMBER OF PROPAGATION CENTERS BY USING RADIOACTIVE QUENCHING AGENTS*

2.1) Some General Remarks on the Method*

The idea to determine the number of propagation centers in olefin polymerization catalysts as the number of the *active* metal-polymer bonds seems to be quite evident. At an early stage of studying the Ziegler-Natta catalysts, this idea was used by Chien(7) for the determination of the number of metal-polymer bonds (MPBs) by quenching polymerization with I_2^{131}. In the subsequent works for the determination of the number of MPBs, radioactive alcohol was used(8,9). In this case the decomposition of MPBs follows the scheme:

$$L_xM \ldots CH_2P + ROH^* \longrightarrow L_xMOR + H^*CH_2P \quad [1]$$

(M - transition metal, L_x - ligands).

---

*We are not going to compare different methods of the determination of the number of propagation centers in Ziegler-Natta catalysts. The data obtained by using different methods are discussed in (10). Here we restrict ourselves only to the results obtained by using radiotracer quenching techniques considering them as the most reliable.

According to the radioactivity of the polymer obtained, the number of propagation centers C_p in the catalyst is calculated:

$$C_p = \frac{A \cdot G}{a \cdot Q} \qquad [2]$$

wherein C_p - the number of propagation centers (moles per mole of transition metal), A - polymer radioactivity (Cu/g), G - polymer yield (g), a - quantity of a catalyst (moles of transition metal), Q - specific radioactivity of a quenching agent (Cu/mole).

However, such a calculation is valid provided that in stopping polymerization by a radioactive quenching agent some evident conditions are met:

(1) The quenching agent should completely stop polymerization.

(2) The quenching agent should interact only with active metal-polymer bonds, not involving inactive ones, i.e. it should be "specific" (see also 2.3 and 2.4).

(3) There should be no radioactive contaminations in the polymer (e.g., as a result of isotopic exchange, side reactions of radioactive compound, adsorption on the catalyst remained in the polymer).

According to the number of propagation centers the propagation rate constant K_p can be calculated:

$$K_p = V/(C_p \cdot C_m) \qquad [3]$$

where K_p - propagation rate constant, l/mole/sec; V - polymerization rate at the moment of introducing a quenching agent, moles of monomer/mole of transition metal/sec; C_m - monomer concentration, mole/l.

The independence of K_p under the variation of those parameters that should not cause the reactivity of propagation centers to change may serve as evidence of K_p quantitative determinations (e.g., such parameters include the change of the absolute catalyst activity value due to the change of the number of propagation centers, the variation of the polymerization duration and of the polymer yield, the use of different types of radioactive quenching agents and the change of their quantity, the monomer concentration variation, etc.).

When catalysts with high activity are used a valid determination of the number of propagation centers by radioactive quenching techniques can be difficult (in particular, in the case of catalytic systems with unsteady activity). However, when the catalyst composition is constant, the K_p value is independent of the polymerization rate, which can vary depending on the art of catalyst preparation and purity of the monomer used for polymerization. The K_p value for the catalysts with the maximum activity (V_{max}) being known, one can calculate the corresponding maximum number of propagation centers ($C_{p.max}$) observed for the given catalytic system. The value $C_{p.max}$ is important when considering various hypotheses on the mechanism of the propagation center formation.

2.2) <u>The Interaction of Various Type Quenching Agents with the Active Centers of Solid Catalysts</u>

TABLE 1.

Incorporation of radioactivity in quenching of ethylene

polymerization.

Catalyst	Quenching agents			
	$C^{14}H_3OH$	CH_3OH^3	$C^{14}O_2$	$C^{14}O$
$TiCl_2$	$-^a$	$+^b$	+	+
CrO_3/SiO_2	+	-	-	+
$Cr(C_3H_5)_3/SiO_2$	-	+	+	+
Catalyst $Cr(C_3H_5)_3/SiO_2{}^c$	+	-	-	+

^{a}+ radioactive polymer. ^{b}- non-radioactive polymer.
^{c}Treated by H_2 at 400°C, free of allyl groups.

Table 1 represents the data on the radioactivity in a polymer produced when ethylene polymerization on different one-component catalysts was stopped by various quenching agents. In the case of stopping polymerization by methanol tagged with tritium in a hydroxyl group, radioactive polymer is obtained with $TiCl_2$ or $Cr(C_3H_5)_3/SiO_2$ as catalysts. This result conforms to the concept of interaction of alcohol with the metal-carbon bond polarized according to the difference of electronegativities of its components (see eq. 2). Earlier the active bond polarization was given much importance, and the catalytic polymerization of olefins was termed as "coordination anionic type" (11).

In the case of the chromium oxide catalyst and of that obtained by the interaction of $Cr(\pi-C_3H_5)_3$ with SiO_2 followed by the elimination of allyl ligands (as a result of catalyst reduction by H_2), upon stopping polymerization by alcohol the alkoxyl group is found in the polymer. These data show that the direction of the interaction of an active metal-polymer bond with alcohol may depend on the active center composition, that seems to influence the type of polarization of the active metal-polymer bond.

Tagged carbon dioxide is incorporated in a polymer molecule in the same manner as the tritium of the hydroxyl group of alcohol. It conforms to the known data on the interaction of CO_2 with organometallic compounds(12):

$$L_xM-CH_2P + C^{14}O_2 \longrightarrow L_xM-O-\overset{\overset{O}{\|}}{C}{}^{14}-CH_2P \quad [4]$$

When stopping polymerization by the introduction of $C^{14}O$, a radioactivity tag was found in the polymer obtained with all the catalysts used (Table 1). After stopping polymerization, as the time of contact of carbon monoxide with the reaction medium increases, slow increase in the number of radioactivity tags in the polymer was observed(13) (see Table 2). For $TiCl_2$ and $Cr(C_3H_5)_3/SiO_2$ upon long contact with $C^{14}O$, the number of tags in the polymer exceeded the number of propagation centers determined in independent experiments with other quenching agents ($C^{14}H_3OH$ or $C^{14}O_2$). The increase in polymer radioactivity does not occur, if, after stopping polymerization by carbon monoxide, phosphine is introduced into the reaction medium. In cases when polymerization is first stopped by phosphine and then $C^{14}O$ is introduced, the polymer will have low radioactivity (see Table 2). These facts show that the insertion of CO into the active metal-polymer bond occurs through a pre-coordination stage:

$$L_xM-CH_2P + C^{14}O \overset{\overset{\overset{C^{14}O}{\downarrow}}{}}{\rightleftarrows} L_xM-CH_2P \longrightarrow L_xM-\overset{\overset{O}{\|}}{C}{}^{14}-CH_2P$$

$$[5]$$

Such a mechanism of the interaction of CO with propagation centers corresponds to the known reactions of insertion of carbon monoxide into the metal-alkyl σ-bond for different organometallic complexes of transition metals(13a). In the case of the systems $TiCl_2$ and $Cr(C_3H_5)_3/SiO_2$ the increase in polymer radioactivity with time in the presence of $C^{14}O$ seems to be caused by slow co-polymerization of $C^{14}O$ with ethylene.

When stopping polymerization by phosphine or carbon monoxide, the number of metal-polymer bonds proved to be the same as in a "working" system (see Table 3). The number of such bonds can be determined by using radioactive alcohol or $C^{14}O_2$. Thus the inhibiting action of such compounds as PH_3

TABLE 2.

Radioactivity in $C^{14}O$ quenched polyethylene as a function of contact time and effect of PH_3.[a]

Catalyst	$C_P \times 10^4$ mole/mole M^b	The number of tags in polymer x 10^4, mole/mole M					PH_3, then $C^{14}O^c$	$C^{14}O$, then PH_3^d
		After treating polymerization medium by $C^{14}O$ contact time, min.						
		5	20	160	360	1200		
$TiCl_2$	0.16	0.08	0.17	0.56	---	---	0.02	0.18
$Cr(C_3H_5)_3/SiO_2$	14	14	---	100	---	650	2.0	13
CrO_3/SiO_2	80	---	11	---	90	70	5.0	95

[a] Polymerization of 75°C (13). [b] M – transition metal; the number of propagation centers was determined in independent experiments using other radioactive quenching agents ($C^{14}H_3OH$, CH_3OH^3, $C^{14}O_2$). [c] The time of contact of $C^{14}O$ with the reaction medium is 360 min. [d] The time of contact of $C^{14}O$ with the reaction medium is 5 min; upon introducing phosphine the mixture of $C^{14}O$ and PH_3 in the case of $TiCl_2$ was kept for 160 min, in the case of $Cr(C_3H_5)_3/SiO_2$ and of CrO_3/SiO_2 for 1200 min.

TABLE 3.

Determination of the number of metal-polymers by different quenching agents.[a]

		Quenching technique		
Catalyst	Polymeriza-tion rate,[b] gC_2H_4/g catalyst/hr	1st quenching agent (non-radio-active)	2nd quenching agent (radio-active)	The number of metal-polymer bonds $\cdot 10^4$, mole/mole Cr
CrO_3/SiO_2	65	no	$C^{14}H_3OH$	28
CrO_3/SiO_2	66	CO	$C^{14}H_3OH$	40
CrO_3/SiO_2	63	PH_3	$C^{14}H_3OH$	36
$Cr(C_3H_5)_3/SiO_2$	140	no	$C^{14}O_2$	14
$Cr(C_3H_5)_3/SiO_2$	136	PH_3	$C^{14}O_2$	15

[a]Polymerization at 75°C (13). [b]At the moment of introducing a quenching agent.

or CO consists primarily in eliminating the vacant coordina-tion site of the transition metal ion of the propagation center:

$$L_xM-CH_2P + X \longrightarrow L_x\overset{X}{M}CH_2P \qquad [6]$$

(X - molecule of a coordination inhibitor strongly bound with the propagation center).

The relative inhibiting action of different compounds depends on the catalyst composition. The results given in Table 4, representing the data on the quantity of inhibitors needed for complete quenching of polymerization, can serve as an example. Carbon monoxide is a stronger inhibitor than CO_2 for the systems $Cr(\pi-C_3H_5)_3/SiO_2$ and $TiCl_2$ (the similar effect was observed in ethylene polymerization by a chromium oxide catalyst where CO is also a stronger inhibitor(14). In the case of the catalysts $Ti(CH_2C_6H_5)_4/Al_2O_3$ and $Zr(\pi-C_3H_5)_4/SiO_2$ (or Al_2O_3), CO_2 is a stronger inhibitor than CO.

TABLE 4.

Inhibition efficiencies[a] of CO and CO_2 in ethylene polymerization by various catalysts.

Catalyst	Polymerization rate, gC_2H_4/mmole M/hr	Inhibitor quantity, mole/mole M[b]	
		CO	CO_2
$Ti(C_7H_7)_4/Al_2O_3$	180	15	2
$Zr(C_3H_5)_4/Al_2O_3$	730	5	0.5
$Zr(C_3H_5)_4/SiO_2$	75	30	5
$Cr(C_3H_5)_3/SiO_2$	600	0.8	100
$TiCl_2$	35	0.02	0.04

[a]Quantity of reagent needed for complete inhibition of polymerization at 80°C, $[C_2H_4]$ = 0.33 mole/l. [b]M – transition metal.

2.3) The Determination of the Number of Propagation Centers in One-Component Catalysts

With relation to the determination of the number of propagation centers by using radiotracer quenching techniques, it is essential that in one-component catalysts there are no inactive non-transition metal-polymer bonds. A priori, it might be expected (the data obtained when introducing phosphine in the polymerization system, given in Table 3, support it) that in polymerization by one-component catalysts two types of transition metal-carbon bonds can be present:

(1) in the active state, being able to undergo the propagation reaction:

$$L_xM\text{-}CH_2P + CH_2 = CH_2 \rightleftharpoons L_xM\text{-}CH_2P \xrightarrow{\quad CH_2=CH_2 \quad}$$

$$L_xM\text{-}CH_2CH_2CH_2P \qquad [7]$$

(2) in the inactive state the centers are deactivated by impurities of the reaction medium (see reaction 6).

However, in the determination of the number of propagation centers in different one-component systems the data does not point to the presence of a measurable number of inactive centers containing a metal-polymer bond (unless a "coordination type" inhibitor was specially introduced into the system). The K_p value remained constant as the polymerization rate varied over a wide range, as a result of the change in purity of the monomer used or in its pressure.

So, it may be assumed that in one-component catalysts all the MPBs are active. To determine their number any tagged quenching agent can be used, the tag of which enters into the polymer chain upon interaction with MPB. As an example the results of the determination of the number of propagation centers for ethylene polymerization by $TiCl_2$ are given in Table 5. In this case different quenching agents ($C^{14}O_2$, propyl iodide tagged by C^{14}, $S^{35}O_2$ and CH_3OH^3) give the same values for K_p.

To determine the number of propagation centers, radioactive carbon monoxide can be also used. To prevent the accumulation of radioactivity in the polymer resulted from copolymerization, phosphine was added after introducing $C^{14}O$ (13). The experimental results show that the same values of K_p can be obtained with short contact of $C^{14}O$ with the reaction medium (for about 5-20 min.); during this period of time only one molecule of CO seems to have inserted into the polymer chain.

The use of C^{14}-tagged quenching agents is preferable, as compared with tritiated ones, due to less probability of isotopic exchange, and in addition, there is no need to take into account the isotopic effect when calculating the values of C_p and K_p. Furthermore, because of the high counting efficiency for C^{14}, it is preferred over other radioactive isotopes when the propagation center concentration is low.

TABLE 5.

Determination of C_p and K_p for ethylene polymerization by $TiCl_2$.[a]

Quenching agent	CH_3OH^3	$C^{14}O_2$	$C_3^{14}H_7I$	$S^{35}O_2$	$C^{14}O$[b]	$C^{14}O + PH_3$
Specific radioactivity of quenching agent, mcuri/mole	2680	2760	452	2230	1210	1210
Concentration of quenching agent, mole/mole Ti	0.14	0.04	0.85	0.02	0.01	0.01 + 0.1
V, $gC_2H_4/gTiCl_2/hr$	305	120	62	165	77	44
$C_p \cdot 10^5$, mole/mole Ti	$3.6 \cdot \alpha$[c]	3.4	2.2	4.5	2.3	1.5
$K_p \cdot 10^{-4}$, 1/mole/sec	$3.1/\alpha$[c]	1.25	1.0	1.35	1.2	1.1

[a] Polymerization at 75°C, $[C_2H_4]$ = 0.33 mole/l. [b] The time of contact of $C^{14}O$ with the reaction medium is 20 min. [c] α – kinetic isotopic effect.

2.4) The Determination of the Number of Propagation Centers in Two-Component Catalysts

In the traditional Ziegler-Natta systems when polymerization proceeds in the presence of an organometallic cocatalyst, chain transfer with an organoaluminum compound takes place(15). As a result, inactive aluminum-polymer bonds are formed that is shown in many works (e.g. 16, 17). In this case the problem of the determination of the number of propagation centers consists in seeking out a "specific" quenching agent, which under the polymerization conditions, interacts with an active transition metal-polymer bond but not with inactive MPBs.

The comparison of the data obtained by using "specific" quenching agents ($C^{14}O$, $C^{14}O_2$) and non-specific ones such as alcohol, is given in Table 6 for the system $TiCl_2+AlEt_2Cl$. This system is convenient, as in this case the data obtained can be compared with the results found for a one-component catalyst - titanium dichloride, where the determination of the number of active bonds can be more unambiguous.

From the data of Table 6 it follows that the number of MPBs found with using radioactive alcohol, exceeds by two orders of magnitude the number of bonds determined with "specific" quenching agents. When CH_3OH^3 was used, the same polymer radioactivity was observed both in the case of a "working" system and when the system was deactivated by the introduction of "specific" quenching agent. These data show that the attempts to calculate the number of propagation centers on the basis of the determination of the total number of MPBs cannot be successful in the case of two-component catalysts.

The similar results were obtained when using different quenching agents in the case of propylene polymerization by the catalytic system $TiCl_3+AlEt_2Cl$ (see Table 7).

As the obtained data show, in the case of two-component systems where diethylaluminum-chloride serves as a cocatalyst, radioactive carbon dioxide can be used to determine the number of active MPBs. It coincides with the known data(12) showing that CO_2 does not interact with organoaluminum compounds, provided that at least one of the alkyl groups is replaced by a halide. $C^{14}O$ can also be used for this purpose taking into account the remarks given above (see 2.3).

When using $C^{14}O_2$ (ethylene polymerization by $TiCl_2$ and $TiCl_2+AlEt_2Cl$), as well as when using successive introduction of $C^{14}O$ and PH_3 (propylene polymerization by $TiCl_3+AlEt_2Cl$) for the determination of the number of propagation centers, the value of K_p did not depend on the quenching agent concentration in the reaction medium, on the absolute number of the

TABLE 6.

Determination of C_p and K_p for ethylene polymerization by the catalyst $TiCl_2 + AlEt_2Cl$.[a]

Quenching agents	$C^{14}O_2$	$C^{14}O$	$C^{14}O+PH_3$	$S^{35}O_2$	CH_3OH^3	CH_3OH^3
[quenching agent], mole/mole Ti	0.045	0.09	0.05	0.11	7.0	7.7
V^b, $gC_2H_4/gTiCl_2/hr$	230	57.5	500	235	---[c]	45
The number of tagged polymer molecules, [mole/mole $TiCl_2$]·10^5	7.0	2.1	16.6	9.2	270α[d]	450α[d]
$K_p \cdot 10^{-4}$, 1/mole/sec	1.3	1.1	1.2	1.0	---	---
τ, min[e]	20	5	5+150	180	30	30

[a] Polymerization at 75°C, Al:Ti=1, $[C_2H_4]$ = 0.3 mole/l. [b] The polymerization rate at the moment of introducing a quenching agent. [c] In this experiment polymerization was stopped first by $C^{14}O$ and then CH_3OH^3 was introduced in the reaction medium. The tritium and C^{14} activities were determined separately. [d] α - kinetic isotopic effect. [e] τ - time of contact of a quenching agent with the reaction medium.

102

TABLE 7.

Determination[a] of C_p and K_p for propylene polymerization catalyzed by $TiCl_3 \cdot 0.3AlCl_3 + AlEt_2Cl$.[b]

Quenching agent	$C^{14}O_2$	$C^{14}O$	$C^{14}O + PH_3$[c]					CH_3OH
[quenching agent], mole/mole Ti	4	0.2	0.1	1.1	0.25	0.5	0.5	6
V, $gC_3H_6/gTiCl_3/hr$	25	51	48	65	6[d]	100[e]	100[e]	71
$C_p \cdot 10^3$, mole/mole Ti	0.36	0.9	0.6	0.74	0.08	1.1	1.1	60α
K_p, 1/mole/sec	70	57	77	90	78	90	90	---

[a]All the data on the polymerization rate refer to the polymer fraction insoluble in the boiling ether. [b]Polymerization at 70°C, $[C_3H_6]$ = 1 mole/l; Al:Ti = 1. [c]The ratio PH_3:Ti = 6, PH_3 was introduced into the reaction medium after 5 min. contact of $C^{14}O$ with the polymerization medium.

103

propagation centers and on the polymer yield. During steady-state polymerization the number of propagation centers did not vary with polymerization time (see, for example, Fig. 1).

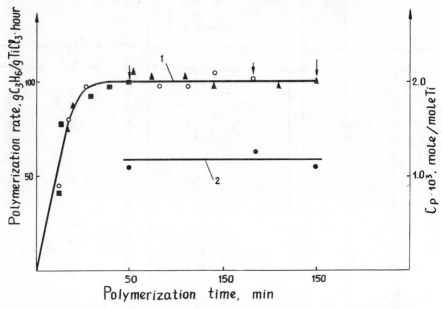

Fig. 1. The variation of polymerization rate (1) and the number of propagation centers (2) with polymerization duration. Catalyst: $TiCl_3 \cdot 0.3AlCl_3 + AlEt_2Cl$, propylene polymerization at monomer concentration 1 mole/1; 70°C, Al:Ti = 1.

3) SUPPORTED OXIDE CATALYSTS

3.1) The Influence of the Catalyst Composition on the Propagation Center Reactivity

The number of propagation centers in oxide catalysts was determined by using $C^{14}H_3OH$ as a radioactive quenching agent (18-21). In the case of chromium oxide catalyst the propagation center reactivity depends on the support (the K_p value varies over the range $Al_2O_3 < ZrO_2 \approx TiO_2 < SiO_2$ (18)) and on the type of reducing agent used for the catalyst treatment (K_p increases over the range $SO_2 < H_2 < NH_3 < HCN$ (21)). The support or the method of catalyst activation being changed, the variation of K_p can be very notable (from ~50 l/mole/sec for CrO_3/Al_2O_3 activated by vacuum treatment, up to ~$3 \cdot 10^3$ l/mole/sec for CrO_3/SiO_2 activated by HCN reduction).

104

The treatment of chromium oxide catalyst with hydrogen-containing reducing agents results in a decrease of the number of propagation centers[21]; this effect might be explained by the decomposition of surface chromium compounds by the water evolved[22]. The treatment of catalysts with hydrogen free reducing agents (SO_2, CO) results in increase of C_p, as compared to the catalyst activated in vacuum (which, in essence, is reduced by ethylene in the reaction medium).

On the basis of the results of the K_p measurements with variation of the composition and the method of activation of the catalyst, it can be concluded that the propagation centers in chromium oxide catalysts have the composition

$$(EO)_x \overset{L}{\underset{|}{Cr}}-CH_2CH_2P$$ (P - polymer chain); their reactivity depends on the element of the support E and the ligand L, coordinated on a chromium ion during the catalyst reduction. The same factors, affecting the active center reactivity in a propagation reaction, are responsible for variations in molecular weight of the polymer obtained as well[23].

Low activity of supported catalysts containing the oxides of Mo, W and V, as compared to chromium oxide catalysts, results from both the low number of propagation centers (C_p in these catalysts did not exceed $\sim 1 \cdot 10^{-3}$ mole per mole of transition metal[24]) and their low reactivity (K_p equals to 5 - 20 l/mole/sec [24]).

The chromium oxide catalyst is a unique oxide polymerization system mainly owing to the possibility of formation of high concentrations of the propagation center precursors (see below).

3.2) The Formation of Propagation Centers in Chromium Oxide Catalysts

On the basis of the data on $C_{p\ max}$ in chromium oxide catalysts some conclusions may be drawn on the mechanism of the propagation center formation. In particular, it is possible to show that propagation centers originate from Cr^{+6} compounds (and not from Cr^{+5} ions).

The hypothesis that the chromium compounds, formed by the interaction of CrO_3 with hydroxyl groups, are the precursors of propagation centers has been proposed long ago[22,25] based on the data concerning the surface composition of chromium oxide catalysts. This hypothesis is supported[26] by the comparison of the maximum number of propagation centers in chromium oxide catalysts with the content of chromium of various oxidation numbers in the catalysts. Such a comparison is given in Table 8. In the catalyst activated by vacuum

TABLE 8.

Comparison of the content of chromium ions in various oxidation states in chromium oxide catalyst (CrO_3/SiO_2) with the maximum number of propagation centers.

	Catalyst activated by heating in vacuum	Catalyst activated by reduction with CO
Catalytic activity at 80°C, $gC_2H_4/$(g catalyst x hour x atm)	160	1000
[Total chromium][a]	$48 \cdot 10^{-5}$	$48 \cdot 10^{-5}$
$[Cr^{+6}]$[a]	$48 \cdot 10^{-5}$	not found
$[Cr^{+\leqslant 3}]$[a]	not found	$48 \cdot 10^{-5}$
$[Cr^{+5}]$[a,b]	$\sim 0.1 \cdot 10^{-5}$	not found
c_p[a]	$6.3 \cdot 10^{-5}$	$17 \cdot 10^{-5}$

[a]mole/g catalyst. [b]Determined by ESR.

treatment, within the accuracy of a chemical analysis, the content of hexavalent chromium coincides with the total chromium content in the catalyst. The number of propagation centers comprises 13% of the total chromium content, that is well above the possible Cr^{+5} content in this catalyst.

In the case of catalysts reduced by carbon monoxide before their interaction with the reaction medium, the number of propagation centers can reach 35% of the total chromium content in the catalyst (see Table 8). According to the data of a chemical analysis, these catalysts do not contain chromium with the oxidation number more than 3. Thus, the propagation center formation in chromium oxide catalysts can be represented by the following scheme:

$\equiv$Si - OH + CrO_3 (or H_2CrO_4) $\longrightarrow$

$\longrightarrow$ [surface compounds Cr^{VI}] $\dfrac{\text{reduction}}{\text{at activation stage or}}$ $\longrightarrow$
in reaction medium

$\longrightarrow$ $\begin{bmatrix}\text{chromium surface com-}\\ \text{pounds with oxidation}\\ \text{number not greater}\\ \text{than 3}\end{bmatrix}$ $\dfrac{\text{alkylation}}{\text{by monomer}} \longrightarrow$

$\longrightarrow$ [Cr - C$\equiv$] $\hspace{4cm}$ [8]

Several arguments, based on the study of catalysts reduced by carbon monoxide by using IR, UV and ESR spectroscopy, can be given to support, that in this scheme the propagation center precursors are surface dichromate ions $(\equiv SiO)_2Cr_2O_5$; these compounds are reduced by carbon monoxide to coordinatively unsaturated Cr^{+3} ions, and it is these ions that undergo alkylation by the monomer(26a).

Up to now no data have been obtained on the mechanism of alkylation of the surface coordinatively unsaturated ions of transition metals by olefins; even though this problem is important for one-component catalysts of various type.

The recently published data(27) on the production of active catalysts by the exchange of the complexes of Cr^{+3} with the surface hydroxyl groups of SiO_2 are consistent with the above scheme. In this case the formation of coordinatively unsaturated low-valent chromium ions proceeds directly upon interaction of the complexes of Cr^{+3} with the support and not upon reduction of Cr^{+6} compounds.

Though many details are still unknown, the current concepts on the active center formation in chromium oxide catalysts can be considered to be satisfactory because it is in agreement with the experimental approaches to active catalysts preparation. If an active catalyst is to be obtained, the maximum concentration of the propagation center precursors should be achieved (the interaction of surface hydroxyl groups with CrO_3 has to be provided, and uncontrolled reduction of Cr^{+6} must be avoided); then their reduction should be carried out with the formation of the maximum number of corrdinatively unsaturated chromium ions (hydrogen free reducing agents have to be applied), and the reactivity of propagation centers should be increased (e.g., by the catalyst treatment with CO). As a result, catalysts having the activity up to 50 kg polyethylene/g Cr/hour/atm can easily be formed.

TABLE 9.

The number of propagation centers and the propagation rate constants for ethylene polymerization by different supported organometallic catalysts.[a]

Catalyst	Quenching agent	Time of contact of quenching agent with reaction medium, min	gC_2H_4/mmole M^b/hr	$C_p \cdot 10^3$ mole/mole M^b	$K_p \cdot 10^{-3}$ 1/mole/sec
Cr(C$_3$H$_5$)$_3$/SiO$_2$[d]	CH$_3$OH$_3$	—	135	0.95α[c]	2.8/α[c]
	C^{14}O	5	130	1.4	1.9
	C^{14}O+PH$_3$	5 for C^{14}O + 1200 for PH$_3$	130	1.5	1.8
Zr(C$_3$H$_5$)$_4$/SiO$_2$	CH$_3$OH$_3$	—	89	15α[c]	0.16/α[c]
	C^{14}O$_2$	—	31	4.1	0.2
	C^{14}O	60	24	6.3	0.1
	C^{14}O+PH$_3$	5 for C^{14}O + 60 for PH$_3$	20	3.0	0.22
Zr(C$_3$H$_5$)$_4$/Al$_2$O$_3$	C^{14}O$_2$	—	890	9.4	2.5
	C^{14}O	60	650	8.7	2.1
	C^{14}O+PH$_3$	5 for C^{14}O + 60 for PH$_3$	760	7.7	2.6
Ti(C$_7$H$_7$)$_4$/Al$_2$O$_3$	C^{14}O$_2$	—	156	3.7	1.11
	C^{14}O	60	265	9.5	0.73
	C^{14}O+PH$_3$	5 for PH$_3$ + 60 for PH$_3$	125	3.6	0.93

[a] Polymerization at 80°, [C$_2$H$_4$] = 0.3 mole/l. [b] M – transition metal. [c] α – kinetic isotopic effect. [d] The experiments at 50°C.

4) *SUPPORTED ORGANOMETALLIC CATALYST*

Upon interaction of π-allyl(3,5), cyclopentadienyl(4), benzyl, neopentyl, trimethylsilyl(5) and some other organometallic derivatives of transition metals with oxide carriers, supported catalysts were obtained showing high activity in ethylene polymerization in the absence of any soluble organometallic cocatalysts. Table 9 represents the data on the number of propagation centers determined for ethylene polymerization by catalysts prepared with using $Cr(\pi-C_3H_5)_3$, $Zr(\pi-C_3H_5)_4$ and $Ti(CH_2C_6H_5)_4$. The number of propagation centers in these catalysts is small, and their reactivity depends both on the type of the transition metal and on that of the support. Low activity of $Zr(\pi-C_3H_5)_4/SiO_2$ catalyst, as compared to that of $Zr(\pi-C_3H_5)_4/Al_2O_3$, results mainly from the decrease in the propagation rate constant when using SiO_2 instead of Al_2O_3 as a support.

The optimum porous structure of the supports and the conditions of the catalysts preparation having been found, the activity of supported organometallic catalysts can be increased. The data on the maximum activity obtained at our laboratory and the results of calculation of the corresponding value of $C_{p\ max}$ are presented in Table 10. In the case of chromium - containing catalysts the maximum number of propagation centers reached 10% of the total content of the transition metal while in the case of zirconium and titanium supported on Al_2O_3 catalysts it did not exceed 2%.

TABLE 10.

Maximum activity for supported organometallic catalysts.[a]

Catalyst	V_{max}, gC_2H_4/mmole $M^{[b]}$/ hr/atm	$C_{p\ max}$, mole/mole M	$K_p \cdot 10^{-3}$, l/mole/sec
$Cr(C_3H_5)_3 + SiO_2$	1900	0.1	2.9
$Zr(C_3H_5)_4 + SiO_2$	65	0.04	0.2
$Zr(C_3H_5)_4 + Al_2O_3$	350	0.02	2.2
$Ti(C_7H_7)_4 + Al_2O_3$	160	0.023	0.95

[a]Polymerization temperature 80°C. [b]M - transition metal.

The interaction of the organometallic compounds of transition elements with the support proceeds essentially with the participation of surface hydroxyl groups:

$$x(E-OH) + MR_n \longrightarrow (EO)_x MR_{n-x} + xRH \quad [9]$$

where E is Si or Al. It should be expected that, depending on the type and the concentration of hydroxyl groups of the support, surface compounds of various types can be obtained. The results of the study of the relationship between the support calcination temperature, determining the concentration of hydroxyl groups on its surface, and the catalytic activity and composition of the complexes formed, are given in Table 11.

From these data it follows, that the number of allyl ligands in the surface compound $(EO)_x MR_{n-x}$ varies with the organometallic compound and the support used and with temperature of the support calcination. The composition of the surface compounds in these catalysts can also vary when they are treated with hydrogen or heated in vacuum (Table 12). It can be seen that catalyst treatment at high temperatures results in complete removal of allyl ligands. At the same time the catalyst retains high activity (in the case of $Zr(C_3H_5)_3/SiO_2$ heating in vacuum results in increased activity). From this it follows, that the presence of allyl ligands bound with a metal ion on the support surface is not compulsory for the propagation centers to be formed.

In the catalysts obtained by the interaction of $Zr(\pi-C_3H_5)_4$ with Al_2O_3 and silica, the ESR signal related to the ions of Zr^{+3} is detected. A certain qualitative correlation between this signal intensity and catalytic activity is observed with variation in temperature of predehydration of alumina and of thermal treatment of the catalyst $Zr(C_3H_5)_4/SiO_2$ in vacuum (Tables 11 and 12). However, such a correlation does not definately implicate the Zr^{+3} ions as the propagation center precursors in these catalysts. The experience (28,29) of the study of chromium oxide catalysts, when comparing the Cr^{+5} content in them with their catalytic activity, show that such correlations do not lead to unambiguous conclusions. Nevertheless, the alkylation of low-valent transition ions by monomers cannot be excluded as one of the possible ways of the propagation center formation in supported organometallic catalysts.

Another possible way of the propagation center formation in these catalysts is ethylene insertion into the bond between metal and organic ligand in the surface complex according to the scheme:

TABLE 11.

The influence of the temperature of calcination of the support on the catalytic activity.[a]

Catalyst	Temp. of calcination of support in vacuum, °C	Metal content on support, mmole M[b]/g	Activity, gC$_2$H$_4$/mmole M/hr.	Molar ratio C$_3$H$_5$: M in catalyst	Relative intensity of ESR signal of Zr^{+3}
Zr(C$_3$H$_5$)$_4$/Al$_2$O$_3$	200	0.51	550	1.2	1
	400	0.40	915	1.4	4.7
	600	0.35	760	1.7	8.2
	800	0.28	540	2.2	3.9
Cr(C$_3$H$_5$)$_3$/SiO$_2$	250	0.86	2000	---	
	400	0.62	6500	1.4	
	700	0.69	215	---	

[a]Polymerization at 80°C, [C$_2$H$_4$] = 0.3 mole/l. [b]M – transition metal. [c]This value corresponds

to 1 to 2% Zr^{+3} of the total zirconium content in the catalyst.

111

TABLE 12.

Effects of hydrogen and heat treatments on the catalytic activity.[a]

Catalyst	Conditions of catalyst treatment	Normalized activity, gC_2H_4/mmole M^b/hr/atm	The number of allyl groups remained in catalyst, mole/mole M^b	Relative intensity of ESR signal of Zr^{+3}
$Cr(C_3H_5)_3/SiO_2$	Heated in vac. at 100°C	33	1.4	
	Treated by H_2 at 30°C	26	0.4	
	Treated by H_2 at 200°C	24	0	
	Treated by H_2 at 400°C	12	0	
$Zr(C_3H_5)_4/SiO_2$	Vacuum at 25°C	15	1.8	1
	Heated in vac. at 100°C	20	1	1.8
	Heated in vac. at 150°C	46	0.36	5.5
	Heated in vac. at 200°C	36	0.05	11[c]
	Heated in vac. at 300°C	20	0	7.7

[a]Polymerization at 80°C. [b]M – transition metal. [c]This value corresponds to 1 to 3% Zr^{+3} of the total zirconium content in the catalyst.

$$(E-O)_x \ M-R_{n-x} + C_2H_4 \longrightarrow (E-O)_x \ M \overset{R_{n-x-1}}{\underset{CH_2CH_2R}{<}} \qquad [10]$$

Such a scheme has been discussed by Ballard *et al.* (5,30). However, this mechanism of the propagation center formation poorly correlates with the low number of propagation centers in the supported zirconium and titanium organometallic catalysts, as well as with the data showing that the samples of supported chromium and zirconium catalysts free of allyl ligands maintain catalytic activity (see Table 12). More unambiguous data on the propagation center precursors in supported organometallic catalysts might be obtained, when using catalysts with the number of propagation centers constituting the major fraction of the surface compounds of the transition metal in the catalyst.

5) *TITANIUM CHLORIDES AS POLYMERIZATION CATALYSTS*

The first step in testing the suitability of a "specific" quenching agent for the determination of the number of propagation centers in polymerization by titanium halides was the study of $TiCl_2$. This catalyst is active in olefin polymerization without aluminum–organic co-catalysts. The comparison of the results obtained with $TiCl_2$ against those for the Ziegler-Natta two-component catalysts and the role of aluminum–organic co-catalysts.

When studying olefin polymerization in the presence of titanium halides with the use of "specific" radioactive quenching agents, the following systems of increasing complexity were studied: $TiCl_2 \longrightarrow TiCl_2$ + aluminum–organic compounds $\longrightarrow TiCl_3$ + aluminum–organic compounds.

5.1) "Monometallic" or "Bimetallic" Active Centers?

In ethylene and propylene polymerization by $TiCl_2$, active centers are sure to be "monometallic" (i.e. their composition involves no metal ions except titanium ones). The introduction of aluminum–organic compounds ($AlEt_3$, $AlEt_2Cl$, $AlEtCl_2$) results in polymerization rate decrease (for ethylene polymerization see, for example, Fig. 2). But experiment shows that the propagation rate constant in ethylene polymerization does not vary when aluminum–organic compounds are added to $TiCl_2$ (compare Tables 5 and 6). The K_p value, determined for ethylene polymerization by $TiCl_3$ + $AlEt_2Cl$ using $C^{14}O$ as a quenching agent was the same ($\sim 1 \cdot 10^4$ 1/mole/sec) as that for polymerization by $TiCl_2$.

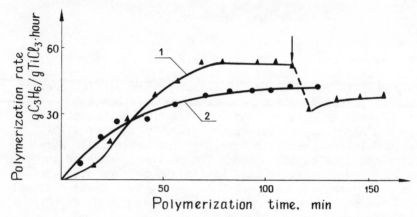

Fig. 2. The influence of introduction of aluminum-organic compounds on ethylene polymerization rate in the presence of TiCl$_2$. Ethylene pressure 2 kg/cm^2; 75°C; 1 - catalyst TiCl$_2$, 2 - catalyst TiCl$_2$ + AlEt$_2$Cl; Al:Ti = 4; an arrow shows the moment of introduction of AlEt$_2$Cl in polymerization with TiCl$_2$ alone.

Table 13 represents the results of the determination of K_p for isotactic addition in propylene polymerization by different catalysts based on titanium chlorides. The polymerization rate being considerably varied, the propagation rate constant is the same (within the accuracy of the experiment) for all the catalytic systems studied. The propagation centers in all these systems seem to have the same composition and are probably "monometallic".

The independence of the K_p value, determined according to the change of molecular weight with polymerization time on the type of an aluminum-organic co-catalyst has been reported earlier(31). However, considerable influence of the type of an organometallic compound on the catalytic system stereospecificity served as the main illustration in favour of the formation of "bimetallic" active centers in two-component catalysts. In this connection it is advisable to consider the data concerning the influence of aluminum-organic compounds on stereoregularity of polypropylene obtained by using TiCl$_2$ (see Table 14), the polymerization rate (Table 15) and the number of propagation centers (Table 16).

TABLE 13.

Number of isotactic propagation centers and the K_p value for propylene polymerization in the presence of different catalysts.[a]

Catalyst	Polymerization rate[b] x 10^3 mole C_3H_6/mole Ti/sec	$C_p \cdot 10^3$ mole/mole Ti	K_p, 1/mole/sec	Quenching agent used in C_p determination
$TiCl_2$	0.099	0.0013	76	$C^{14}O_2$
$TiCl_2$+AlEt$_2$Cl	0.080	0.00085	94	$C^{14}O_2$
$TiCl_3$[c]+AlEt$_2$Cl	1.52	0.025	61	$C^{14}O$ + PH_3
$TiCl_3 \cdot 0.3AlCl_3$[d]+AlEt$_2$Cl	102	1.4	73	$C^{14}O$ + PH_3
$TiCl_3 \cdot 0.3AlCl_3$[d]+AlEt$_3$	192	2.4	80	$C^{14}O$ + PH_3

[a]Polymerization at 70°C; Al:Ti = 1. [b]The rate of isotactic polypropylene formation normalized to the monomer concentration 1 mole/l. [c]A sample of TiCl$_3$ was obtained by hydrogen reduction of TiCl$_4$ at 800°C. [d]A sample of TiCl$_3 \cdot 0.3AlCl_3$ was obtained by aluminum reduction of TiCl$_4$ and activated by dry grinding.

TABLE 14.

Properties of polypropylene obtained on different catalysts.[a]

Catalyst	Fraction soluble in ether[b]		Fraction soluble in boiling heptane[c]		Fraction insoluble in boiling heptane[d]	
	% wt	$[\eta]$,[e] dl/g	% wt	$[\eta]$,[e] dl/g	% wt	$[\eta]$,[e] dl/g
$TiCl_2$	60	2.6	15	4.2	25	12
$TiCl_2+AlEt_2Cl$	30	1.6	15	3.8	55	8.6
$TiCl_2+P(C_6H_5)_3$	28	3.1	12	5.0	60	8.5
$TiCl_3 \cdot 0.3AlCl_3$ $+AlEt_2Cl$	2	0.9	2	2.2	96	11.5

[a]Polymerization at $70°C$ in liquid propylene in the case of $TiCl_2$, and in gasoline at monomer concentration 1 mole/1 in the case of $TiCl_3$; Al:Ti = 1; $P(C_6H_5)_3$:Ti = 1.20. [b]Amorphous according to X-ray method (32). [c]Medium crystallinity. [d]High crystallinity. [e]Intrinsic viscosity (decalin, $135°C$).

TABLE 15.

The influence of aluminum-organic compounds on the polymerization rate and isotacticity of polypropylene obtained on $TiCl_2$.[a]

Aluminum-organic compounds	None	$AlEt_3$	$AlEt_2Cl$	$AlEtCl_2$
Polymerization rate, $gC_3H_6/gTiCl_2$/hour	5.6	4.6	2.2	0.2
Fraction insoluble in boiling n-heptane, % weight	25	70	55	22

[a]Polymerization in liquid propylene at $70°C$; Al:Ti = 1.

TABLE 16.

Value of C_p and K_p determined by tracer method for various polymer fractions.[a]

	Fraction soluble in ether		Fraction soluble in boiling heptane		Fraction insoluble in boiling heptane	
	C_p	K_p	C_p	K_p	C_p	K_p
$TiCl_2$ (average polymerization rate 5.6 $gC_3H_6/gTiCl_2/hour$)	36	63	4.5	126	12.6	76
$TiCl_2$ + $AlEt_2Cl$ (average polymerization rate 2.2 $gC_3H_6/gTiCl_2/hour$)	5.5	80	2.5	87	8.5	94

[a]Polymerization in liquid propylene at 70°C; Al:Ti = 1; quenched by $C^{14}O_2$.

According to the extent of inhibiting action on propylene polymerization, aluminum-organic compounds are arranged in the series $AlEt_3 < AlEt_2Cl < AlEtCl_2$. Approximately 25% of propylene obtained in polymerization by $TiCl_2$ does not dissolve in boiling *n*-heptane (an isotactic fraction). Upon introducing aluminum-organic compounds, the amount of this fraction increases. The increase in the catalyst stereospecificity is observed in the series $AlEtCl_2 < AlEt_2Cl < AlEt_3$.

The tracer data on the number of propagation centers in various fractions of polypropylene obtained on the catalysts $TiCl_2$ and $TiCl_2 + AlEt_2Cl_2$ (Table 16), enables us to explain these results. With the addition of $AlEtCl_2$ to $TiCl_2$ there is a sharp reduction in the number of nonstereospecific active centers, the number of stereospecific active centers varies only slightly. The value of the propagation rate constant is not significantly affected by the addition of an aluminum-organic compound to $TiCl_2$; the observed fall of the polymerization rate is attributed mainly to the decrease in the number of nonstereospecific active centers.

Our experimental data are in agreement with the presence
of monometallic active centers of dissimilar stereospecificity
on the surface of $TiCl_2$; the nature of an active center seems
to be determined by the number and the arrangement of chlorine
ions on the titanium (as it was assumed by Cossee and Arlman
for polymerization by $TiCl_3$(33-35). Aluminum-organic compound
is adsorbed on the active centers, e.g., according to the
scheme:

$$ \text{(A)} + AlR_3 \rightleftarrows \text{(B)} \qquad [11] $$

($\square$ - vacant coordination site, P - polymer chain). In this
case the active state A of the propagation center changes to
an inactive one (B) on which no growth of the polymer chain
occurs. As the adsorption of aluminum-organic compounds is
selective with respect to active centers of different specificity, the ratio between the number of "working" active
centers of various stereoregularity types varies with addition
of aluminum-organic compounds and causes the polymer stereoregularity to change. Thus, the addition of any compounds
capable for selective adsorption on various active centers
should affect activity and stereospecificity of the catalyst
system. For example, in propylene polymerization by titanium
dichloride the introduction of triphenylphosphine results in
sharp increase in the catalyst stereospecificity, approaching
the case of $TiCl_2$+$AlEt_2Cl$ (see Table 14). The variation of
the ratio between the number of atactic and isotactic active
centers as influenced by an aluminum-organic compound seems
to be determined by the size of the molecule and the mode of
its interaction with an active center. The isotactic active
centers are less sterically available than the atactic ones,
therefore, as the molecule size diminishes (e.g., when $AlEt_3$
changes to $AlEt_2Cl$), one can expect the proportion of isotactic propagation centers in the active state to decrease
and, thus, the polymer isotacticity to diminish (see Tables
14 and 15).

So, the influence of organometallic compounds on stereospecificity of polypropylene, obtained by using $TiCl_2$, is
attributed to variation of the ratio between the number of
isotactic and atactic active centers. That the two types of
the centers are "monometallic" is supported by the fact that

118

the rate constant of chain growth is the same for both (see Table 16), and within the experimental accuracy is coincident with the K_p value found in the absence of organometallic compounds. An analogous mechanism of the influence of organometallic compounds on the catalyst stereospecificity seems to be also valid for the systems based on $TiCl_3$. But in the case of the two-component systems "$TiCl_3$ + an organometallic co-catalyst" the influence of the latter is two-fold:

(1) in the process of alkylation of the surface titanium ions (the initial stage) an organometallic compound can influence the ratio between the formed centers of various stereospecificity.

(2) in the process of polymerization the selective adsorption of an organometallic compound on the centers of various stereospecificity can determine the overall rate and polymer stereoregularity.

The participation of aluminum-organic co-catalysts in these two processes may determine that complicated relationship between the stereospecificity of two-component systems and the type of an organometallic compound, that was observed for various transition metal halides used as olefin polymerization catalysts(36).

5.2) The Proportion of "Working" Surface Titanium Ions

In the formation of a crystalline polymer on the surface of a solid catalyst the concept of a "working" surface of the catalyst is not evident. As a rule, no correlation between the catalytic activity and the BET surface area of $TiCl_2(1,37)$ or $TiCl_3(38,39)$ is observed. It can be attributed to fragmentation of the catalyst during polymerization uncovering fresh catalyst surface.

The correlation between the activity of $TiCl_2$ samples in propylene polymerization and their dispersity, determined by using X-ray techniques, have been observed recently(40); the dispersity thus obtained may characterize the size of the primary crystallites composing a macroparticle of the catalyst. The analogous correlation was found by us when comparing the activity of several samples, prepared by $TiCl_4$ reduction by organoaluminum compounds, with their dispersity (see Fig. 3).

When calculating the surface concentration of propagation centers based on radiotracer quenching data, the proper reference point should be the specific surface in a "working" catalyst (after fragmentation by the polymerization). With certain limitations this specific surface might be calculated from the data on the dispersity of $TiCl_3$ crystallites obtained by X-ray techniques. Using such a value of specific surface

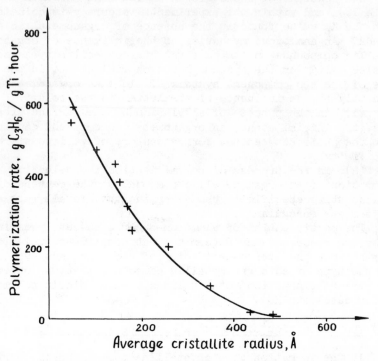

*Fig. 3. The dependence of activity of propylene polymeriza-
tion catalysts of the composition $TiCl_3 \cdot n \, AlCl_3$ ($n = 0.2$
$+0.5$), obtained by using various methods, on their dispersion. The dispersion of the samples was determined
by using X-ray techniques as the size of primary crystallites along the <001> axis. Polymerization at 70°C,
cocatalyst $AlEt_2Cl$, monomer concentration 1 mole/l.*

we obtain a linear relationship between this value and the
catalyst activity (see Fig. 4).

Table 17 presents the data on the maximum activity,
observed in ethylene polymerization with $TiCl_2$ and in propylene polymerization with $TiCl_3$, and the corresponding values
of C_p max.

TABLE 17.

Maximum activity of $TiCl_2$ and $TiCl_3$ in olefin polymerization and the corresponding number of propagation centers.

Catalyst	Specific surface area, m^2/g	Olefin	V_{max}, g olefin/ g Ti/hr [a]	$C_{p,\ max.}$ mole/mole Ti	$C_{p,\ max.}$ mole/mole of surface Ti
$TiCl_2$	>25[b]	ethylene	3500	$1.5 \cdot 10^{-4}$	<0.004
$TiCl_3 \cdot 0.3AlCl_3$ + $AlEt_3$	250[c]	propylene	4000	$1.3 \cdot 10^{-2}$	0.025

[a]The activity at 70°C is normalized to a monomer concentration of 1 mole/l. [b]Determined by the BET method. [c]The specific surface of crystallites calculated from its average size as determined by x-ray techniques (see also Fig. 4).

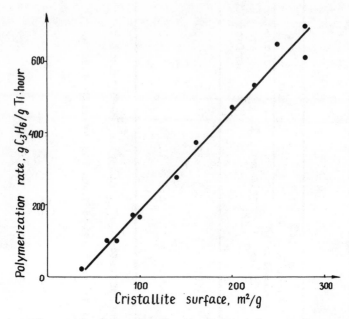

*Fig. 4. The ratio between the activity of propylene polymer-
ization catalysts and their specific surface. The
specific surface of the samples was calculated according
to average sizes of primary crystallites determined by
using X-ray techniques on the assumption that the crystallites have the form of hexagonal prisms. The polymerization conditions are given in the caption to Fig. 3.*

In the case of $TiCl_2$ the number of propagation centers
is very small, for the investigated samples $C_{p\ max}$ reached
$1.5 \cdot 10^{-4}$ mole/mole Ti. The surface concentration of the propagation centers does not exceed 0.4% of the number of titanium ions adjacent to the surface of the initial catalyst prior
to its interaction with the reaction medium. Taking into
account the possibility of the breaking up of the catalyst to
increase the "working" surface, the surface concentration of
propagation centers should be assumed to be still lower.
Such a small number of propagation centers seems to be determined by the fact that ethylene alkylates only titanium ions
having high coordination insufficiency, the proportion of
which is low.

As it follows from the data of Table 17, in polymerization with $TiCl_3$ the maximum number of propagation centers
reached the value $1.3 \cdot 10^{-2}$ mole/mole Ti. On the basis of BET
surface area ($\sim 25 m /g$) one might erroneously conclude that

the surface titanium ions are intensively used in the formation of active centers (such conclusions were drawn, e.g.,
(41-43)). However, we believe that this number should be
related to a "working" surface, which is ca. 250 m^2/g by X-
ray, then the proportion of propagation centers equals only
to about 2.5% of the number of surface titanium ions.

We believe that the results of the determination of propagation centers by using radiotracer quenching techniques
provide evidence in favour of that point of view, according
to which the formation of propagation centers in the case of
unsupported $TiCl_3$ proceeds with the participation of only
those surface titanium ions that are situated in special surface regions (e.g., on the outcrops of growth spirals, on
lateral faces, etc.(34,44)).

Of interest is the question on the proportion of "working"
titanium ions in supported two-component polymerization catalysts. That catalysts, obtained by depositing $TiCl_4$ onto
various supports with the subsequent reduction of Ti^{+4} to
Ti^{+3} (in the process of catalyst preparation or in the reaction medium by an organometallic co-catalyst), are highly
active in ethylene polymerization (6). Table 18 represents
the data obtained in our Laboratory on the activity of supported catalysts prepared by using different supports.

TABLE 18.

Polymerization rate and number of propagation centers for

ethylene polymerization on supported titanium catalysts.

Support, used for catalyst preparation	V_{max}, [a] $gC_2H_4/gTi/hr/atm$	$C_{p, max}$, [b] mole/mole Ti
Polyethylene	18,000	0.013
Silica	15,000	0.011
Magnesium compounds	200,000	0.15

[a]Polymerization at 80°C. [b]$C_{p, max}$ calculated with $K_p = 1 \cdot 10^4$
1/mole/sec for $TiCl_3$ as determined by $C^{14}O$ quenching.

When using powdered polyethylene as a support, the increase in dispersity of $TiCl_3$ phase in the support pores is obviously responsible for the increase in the catalyst activity. It is confirmed by the results of the determination of the propagation rate constant for ethylene polymerization. When using $C^{14}O$ as a radioactive quenching agent, the K_p value in ethylene polymerization by this catalyst is the same as it was found in the case of $TiCl_2$ and $TiCl_3$ ($1\cdot10^4$ 1/mole/sec at 80°C). The data obtained when studying the composition of the catalysts, containing titanium chlorides on silica(45), and the kinetics of ethylene polymerization by these catalysts(45a) show, that in this case the $TiCl_3$ microphase is an active component of the catalyst as well. Using the value $K_p = 1\cdot10^4$ 1/mole/sec the proportion of "working" titanium ions in supported catalysts can be calculated (Table 18). It equals to about 1% of the total number of titanium ions in catalysts on polyethylene and silica.

The catalysts obtained by using various magnesium compounds as a support have still higher activity. As yet for these catalysts the K_p value has not been determined, but if the K_p value is assumed to remain the same as for polymerization with $TiCl_3$ deposited onto polyethylene, then the proportion of the propagation centers in such catalysts will reach about 15% of the total titanium content in the catalyst. Thus, all the surface titanium ions could be used here. However, it may be possible that in this case the propagation center composition will differ from that observed when using crystalline $TiCl_3$ as a catalyst, that results in different reactivity in a propagation reaction. The cause of high activity of these catalysts (drastic increase in dispersity of $TiCl_3$ phase or formation of new-type propagation centers having high reactivity upon interaction of $TiCl_4$ with a support?) will be clarified after the determination of the number of propagation centers and their reactivity for these catalysts by using "specific" radioactive quenching agents.

4.3) The Determination of the Chain Transfer Rate Constants

The K_p value may be used for the calculation of the rate constants of chain transfer processes and the time of polymer molecule growth according to the data on the molecular weight of polymer. In ethylene polymerization by $TiCl_2$ the transfer with a monomer is predominant (the polymer molecular weight does not vary with the increase in ethylene pressure above 2 kg/cm^2). In this case the number average degree of polymerization $\overline{P}$ is determined as:

$$\bar{P}_n = \bar{P}_w/\gamma = K_p/K_m \qquad [12]$$

wherein $\bar{P}_w$ - weight average degree of polymerization;

$\gamma = \dfrac{\bar{P}_w}{\bar{P}_n}$ - parameter of polymer polydispersity; K_m - rate constant of transfer with a monomer.

The mean time of the polymer molecule growth can be calculated:

$$\bar{\tau} = \frac{\bar{P}_n}{K_p \cdot C_m} = \frac{\bar{P}_w}{\gamma \cdot K_p \cdot C_m} \qquad [13]$$

wherein C_m - monomer concentration.

The calculated values K_p and τ with an accuracy of the parameter γ are given in Table 19 (a "radiochemical" method).

TABLE 19.

Rate constants for propagation (K_p) and transfer to monomer (K_m) and average time for polymer growth (τ) in ethylene polymerization. [a]

Method of determination	$K_p \cdot 10^{-2}$, l/mole/sec	$K_m \cdot 10^2$, l/mole/sec	τ, [b] sec
Radiochemical	200	$30 \cdot \gamma$ [c]	$18/\gamma$
Kinetic	$3 \cdot 6/\gamma$	0.5	1000

[a] $TiCl_2$ catalyzed polymerization at 90°C. [b] Calculated at ethylene concentration 0.2 mole/l. [c] γ - polydispersity parameter.

First, this method was used by Natta *et al.* (15,46), and it was often used in the analysis of the kinetics of polymerization by the Ziegler-Natta catalysts(31,43,47,48). When the polymerization rate varies with time, for the calculation of the K_p value the following equation can be used(49):

$$1/\bar{P}_w = \frac{1}{\gamma \cdot K_p \cdot C_m} \cdot V/G + \frac{K_m}{\gamma \cdot K_p} + \frac{K_c}{\gamma \cdot K_p \cdot C_m} \qquad [14]$$

wherein V - polymerization rate at the moment τ; G - polymer yield at the same moment, K_c - rate constant of spontaneous chain transfer.

According to the polymer molecular weight obtained over the region of monomer concentrations where $\overline{P}_w$ does not depend on C_m, the K_m value can be calculated:

$$\overline{P}_w = \frac{\gamma K_p}{K_m} \qquad [15]$$

TABLE 20

Molecular weight of polyethylene.[a]

Polymerization time, min	Ethylene concentration, mole/l	$M_w \cdot 10^{-6}$ [b]
7	0.20	0.37
15	0.20	0.9
30	0.20	1.5
87	0.20	1.7
130	0.10	1.8
87	0.20	1.7
86	0.35	1.4
45	1.25	1.8

[a]$TiCl_2$ catalyzed polymerization at 90°C. [b]Calculated according to (50):$[\eta] = 2.55 \cdot 10^{-4} \cdot \overline{M}_w^{0.74}$. The intrinsic viscosity $[\eta]$ was determined in decalin at 130°C.

Table 20 presents the data on the molecular weight of polyethylene obtained under different conditions, on the basis of which the γK_p and K_m values were calculated by using a "kinetic" method. The relationship in coordinates corresponding to eq. 14 is given in Fig. 5. The values of rate constants of separate stages obtained by using a "kinetic" method for ethylene polymerization with $TiCl_2$ are close to the values of the corresponding constants, found by using the same method for ethylene polymerization by the system $TiCl_3+AlEt_2Cl$ (47). However, these data differ greatly from the values found by using a "radiochemical" method.

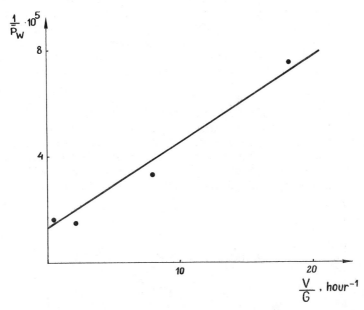

Fig. 5. The relationship in coordinates $1/\overline{P}_w$ versus V/G for ethylene polymerization with $TiCl_2$ at 90°C and monomer concentration 0.2 mole/l; V - polymerization rate at the moment τ; G - polymer yield for the same time.

The discrepancy between the data on the reactivity of propagation centers obtained by using two methods is too large to be reconciled. When analyzing the relationship between the molecular weight and the polymerization time, several factors which might affect this relationship seems to not have been taken into account. In the case of one-component catalysts the most probable cause of molecular weight variations in early stages of polymerization, not taken into account, is chain transfer with the participation of impurities. In the course of polymerization the microquantity content of impurities can decrease as a result of interaction with propagation centers and the surface of catalysts, that may affect the relationship between the molecular weight and τ, and result in variation of polydispersity of the polymer obtained.

For two-component catalyst chain transfer with an aluminum alkyl co-catalyst results in the formation of aluminum-polymer bonds. The number of these bonds (N_{Al-P}) is determined upon stopping polymerization by tagged alcohol. According to the data of Table 6 for ethylene polymerization by $TiCl_2 + AlEt_2Cl$ the ratio between N_{Al-P} and the total number of the polymer molecules formed (N_{tot}) is estimated to be $N_{Al-P}/N_{tot} = 1.8(\alpha/\gamma)$, wherein α is the kinetic isotopic effect, and γ is the polydispersity parameter. Generally α = 1 to 4, γ = 3 to 10. Thus, the proportion of polymer molecules, formed as a result of transfer with an aluminum alkyls, is not less than 20% of the total number of the polymer molecules. A similar estimate for propylene polymerization cata-lized by $TiCl_3 \cdot 0.3$ $AlEtCl_2$ (see the data on the number of aluminum-polymer bonds in Table 7) shows that the transfer with aluminum alkyl co-catalyst results in the formation of not less than 30% of the polymer molecules. It is natural to assume the reaction of transfer with an aluminum alkyls to proceed via adsorption of the latter on a titanium ion in the propagation center(51):

$$[15]$$

The temporary deactivation of propagation centers by adsorption of aluminum alkyl compounds may cause a step-wise process of formation of the polymer molecule: after a rapid reaction of the monomer insertion into an active bond the propagation center goes into the inactive state, the polymer chain remains bound to the propagation center. Then the aluminum alkyl desorption occurs followed by the insertion of the monomer, etc. In this case "life-time" of the polymer chain can reach a reasonable value; in several papers the increase in polypropylene molecular weight was observed for several hours(52,53). However, during this period the poly-mer chain is essentially inactive, and the proper time of its growth determined by eq. 12 is small. For propylene polymer-ization by $TiCl_3 + AlEt_2Cl$ it equals to the value not longer than 3 min (70°C, monomer concentration 1 mole/l).

6) *CONCLUSION*

The data on the number of propagation centers in olefin polymerization catalysts are of great importance serving as a criterion for the discrimination of alternative hypotheses on the mechanism of their action. These data along with those concerning reactivity of propagation centers in separate stages of polymerization processes also are useful to chemists concerned with the development of new polymerization catalysts and methods of regulating polymer properties. In addition, a knowledge of the rate constants of separate stages and the corresponding activation energies enables us to understand the relationship between the composition of active centers and their reactivity.

Many aspects of the mechanism of action of solid olefin polymerization catalysts have not yet been solved. In particular, there is no data on the mechanism of alkylation of low-valent transition metal ions by a monomer resulting in formation of an active metal-carbon bond in one-component catalysts. The important problem for the catalysts of all types consists of the elucidation of the detailed mechanism of various reactions of chain transfer along with the determination of rate constants of separate steps composing this complicated process. The determination of the relationship between the composition of propagation centers and their reactivity remains a key problem of catalytic polymerization. To solve this problem catalysts are needed which have the same number of propagation centers as that of the transition metal on the surface (or, at least, to its predominant proportion). In this case to obtain the data on the composition of propagation centers different physical methods might be used.

We believe that the solution of these problems may be facilitated by the study of supported organometallic catalysts. In principal, the method of preparation of supported catalysts by using organometallic compounds of transition elements may permit the type of a transition metal and its ligand environment in the propagation center to be widely changed. However, for the catalysts of this type the methods of their preparation, when the number of propagation centers is close to the content of a supported transition metal, are still to be developed.

The method of preparation of supported catalysts with the use of organometallic compounds of transition elements, described at the beginning(3-5) as the method of preparation of polymerization catalysts, proved to be useful for the preparation of catalysts for non-polymerization processes as well. The surface organometallic compounds of transition elements obtained by using such a method, proved to be active in olefin disproportionation(54). By the reduction of these compounds, either low-valent ions of transition metal active in hydrogenation of olefins(55) and nitrogen(56), or particles of superdispersed metals(57-59) can be obtained. By the oxidation of surface organometallic complexes, surface compounds can be obtained containing high-valent transition metals showing interesting properties in oxidation reactions(60).

The urgent problem in the field of coordination catalysts – polymerization as well as non-polymerization – consists in the development of the methods of synthesis of active centers as surface compounds of transition metals with certain ligand environment and necessary coordinative unsaturation. The solution of this problem should lead to the development of "precise" catalysts, all the transition metal ions in which will be involved in the composition of active centers, and all the active centers will show the same catalytic properties.

ACKNOWLEDGEMENT

The authors wish to acknowledge the help of all their colleagues working at the polymerization laboratory at the Institute of Catalysis submitted the data considered in this review: G.D. Bukatov (the determination of the number of propagation centers by using radiotracer quenching techniques; the study of polymerization with $TiCl_2$), M.B. Chumaevski (the study of polymerization with $TiCl_3$), E.G. Kushnareva (the investigation of chromium oxide catalysts), E.E. Vermel, P.A. Zhdan (the study of $TiCl_3$), V.K. Dudchenko, A.I. Min'kov, O.N. Efimov and E.A. Demin (the study of supported organometallic catalysts). The authors are also grateful to G.D. Bukatov for the discussion of the problems of polymerization mechanism and a number of valuable remarks on the content of this review.

LITERATURE CITED

(1) C.X. Werber, C.J. Benning, W.R. Wszolek and G.E. Ashby, *J. Polymer Sci.*, *A-1*, *6*, 743 (1968).

(2) A.S. Matlack and D.S. Breslow, *J. Polymer Sci.*, *A*, *3*, 2853 (1965).

(3) Yu.I. Yermakov, A.M. Lazutkin, E.A. Demin, V.A. Zakharov and Yu.P. Grabovski, *Kinetika i kataliz*, *13*, 1422 (1972).

(4) F.J. Karol, G.L. Karapinka, Ch. Wu, H.W. Dow, R.N. Johnson and W.L. Carrick, *J. Polymer Sci.*, *A-1*, *10*, 2621 (1972).

(5) D.G.H. Ballard, *Advan. Catalysis*, *23*, 267 (1973).

(6) *Hydrocarbon Process.*, *49*, Nll, 179 (1970); *ibid.*, *51*, Nll, 130 (1972); *ibid.*, *51*, N7, 115 (1972); *Chem. Eng.*, *79*, April 3, 66 (1972).

(7) J.C.W. Chien, *J. Amer. Chem. Soc.*, *81*, 86 (1959).

(8) C.F. Feldman and E. Perry, *J. Polymer Sci.*, *46*, 217 (1960).

(9) J.C.W. Chien, *J. Polymer Sci.*, A-1, 425 (1963).

(10) Yu.I. Yermakov and V.A. Zahkarov, *Uspekhi khimii*, *41*, 377 (1972).

(11) G. Natta, F. Danusso and D. Sianesi, *Makromol. Chem.*, *30*, 238 (1959).

(12) Organometallic Chemistry edited by Zeiss, Reinhold Publishing Corporation, New York, 1960).

(13) V.A. Zakharov, G.D. Bukatov, Yu.I. Yermakov and E.A. Demin, *Doklady Akademii Nauk*, *207*, 857 (1972).

(13a) J.P. Candlin, K.A. Taylor and D.T. Thompson, "Reactions of Transition Metal Complexes", Elsevier Publishing Co., 1968.

(14) Yu.I. Yermakov, L.P. Ivanov and A.I. Gelbshtein, *Kinetika i kataliz*, *10*, 411 (1969).

(15) G. Natta and I. Pasquon, *Advances in Catalysis*, *11*, 1 (1959).

(16) E. Kohn, H. Shuurmans, J.V. Cavender and R.A. Mendeleson, *J. Polymer Sci.*, *58*, 681 (1962).

(17) H.W. Coover, J. Guillet, R. Combs and F.B. Joyner, *J. Polymer Sci.*, A-1, *4*, 2583 (1966).

(18) Yu.I. Yermakov, V.A. Zakharov, The Fourth International Congress on Catalysis, Moscow, 1968, Preprint 16; Trudy IV Mezhdunarodnogo Kongressa po Katalizu, Nauka, Moscow 1970, p. 200.

(19) V.A. Zakharov and Yu.I. Yermakov, *J. Polymer Sci.*, A-1, *9*, 3129 (1971).

(20) Yu.I. Yermakov, V.A. Zakharov and E.G. Kushnareva, *J. Polymer Sci.*, A-1, *9*, 771 (1971).

(21) E.G. Kushnareva, Yu.I. Yermakov and V.A. Zakharov, *Kinetika i Kataliz*, *12*, 414 (1971).

(22) J. Hogan, *J. Polymer Sci.*, A-1, *8*, 2637 (1970).

(23) Yu.I. Yermakov, L.P. Ivanov, E.G. Kushnareva, V.A. Zakharov and A.I. Gelbshtein, *Kinetika i Kataliz*, *10*, 590 (1969).

(24) V.A. Zakharov, E.G. Kushnareva and Yu.I. Yermakov, *Kinetika i Kataliz*, *10*, 1164 (1969).

(25) T. Hill and J. Habeshaw, in Proceedings of the 3rd International Congress on Catalysis, Amsterdam, 1964, Amsterdam, 1965, p. 975.

(26) Yu.I. Yermakov, V.A. Zakharov, Yu.P. Grabovski and E.G. Kushnareva, *Kinetika i Kataliz*, *11*, 519 (1970).

(26a) V.A. Zakharov and E.G. Kushnareva *et al.*, *Kinetika i Kataliz*, in press.

(27) Z.K. Przhevalskaya, V.A. Shvetz and V.B. Kazanski, *Kinetika i Kataliz*, *15*, 2 (1974).

(28) G.K. Boreskov, F.M. Bukanaeva, V.A. Dzisko, V.B. Kazanski, and Yu.I. Petcherskaya, *Kinetika i Kataliz*, *5*, 434 (1964).

(29) L.Ya. Alt, V.F. Anufrienko, T.Ja. Tyulikova and Yu.I. Yermakov, *Kinetika i Kataliz*, *9*, 1253 (1968).

(30) P.G.H. Ballard and P.W. Van Lienden, *Makromolec. Chemie*, *154*, 177 (1972).

(31) G. Natta, A. Zambelli, I. Pasquon and G.M. Giongo, *Chim. e Ind.*, (Milan), *48*, 1298 (1966).

(32) G. Natta, P. Corradini and M. Cesari, *Atti Accad. Naz. Lincei*, *Ren.*, *22*, 11 (1957).

(33) P. Cossee, *J. Catalysis*, *3*, 80 (1964).

(34) E.I. Arlman, *J. Catalysis*, *3*, 89 (1964).

(35) E.I. Arlman and P. Cossee, *J. Catalysis*, *3*, 99 (1964).

(36) G. Natta, A. Zambelli, I. Pasquon and G.M. Giongo, *Chim. e Ind.*, (Milan), *48*, 1307 (1966).

(37) G.D. Bukatov, V.A. Zakharov and Yu.I. Yermakov, *Kinetika i Kataliz*, *12*, 505 (1971).

(38) T. Keii, "Kinetic of Ziegler Natta Polymerization", Kodansha, Tokyo, 1972.

(39) Ja. Ambroz, R. Vespalec and O. Hamrik, *Chem. oný pzůmysl*, *18*, 549 (1968).

(40) Z.W. Wilchinsky, R.W. Looney and E.G.M. Tornqvist, *J. Catalysis*, *28*, 351 (1973).

(41) Yu.V. Kissin, S.N. Mezhikovsky and N.M. Chirkov, *Eur. Polymer J.*, *6*, 267 (1970).

(42) L.A. Novokshonova, N.M. Chirkov and V.I. Tsvetkova, Kinetics and Mechanism of Polyreactions, Preprints, Intern. Sypmos. on Macromolecular Chem., *Akad.*, Kiado, Budapest, 1969, Vol. 2, p. 249.

(43) V. Murayama and T. Keii, *Shokubai (Catalyst)*, *5*, 247 (1963).

(44) L.A.M. Rodriguez, H.M. Van-Looy and I.A. Gabant, *J. Polymer Sci.*, *A-1*, *4*, 1971 (1966).

(45) N.G. Maximov, E.G. Kushnareva, V.A. Zakharov, V.F. Anufrienko, P.A. Zhdan and Yu.I. Yermakov, *Kinetika i Kataliz*, *15*, N3, 1974.

(45a) V.A. Zakharov, V.N. Druzhkov, E.G. Kushnareva and Yu.I. Yermakov, *Kinetika i Kataliz*, *15*, 443 (1974).

(46) G. Natta, I. Pasquon, J. Svab and A. Zambelli, *Chim. e Ind.*, *44*, 621 (1962).

(47) B.M. Grievson, *Makromol. Chem.*, *84*, 93 (1965).

(48) L.A. Novokshonova, G.P. Berseneva, V.I. Tsvetkova and
      N.M. Chirkov, *Vysokomol. soed., A-9,* 562 (1967).
(49) Yu.I. Yermakov, L.P. Ivanov and A.I. Gelbshtein,
      Kinetika i Kataliz, 10, 184 (1969).
(50) M.O. De la Cuesta, I.W. Billmeyer, *J. Polymer Sci., A-1,
      1,* N5, 1721 (1963).
(51) P. Cossee, *Trans. Farad. Soc., 58,* 1226 (1962).
(52) G. Bier, W. Hoffmann, G. Lehman and G. Seydel, *Makromol.
      Chem., 58,* 1 (1962).
(53) A.K. Ingberman, I.J. Levine and R.J. Turbett, *J. Polymer
      Sci., A-1, 4,* 2781 (1966).
(54) Yu.I. Yermakov, B.N. Kuznetsov and A.N. Startsov, *Kine-
      tika i Kataliz, 15,* 556 (1974).
(55) Yu.I. Yermakov and B.N. Kuznetsov, Preprints of the
      Second Soviet - Japanese Seminar on Catalysis, Tokyo,
      October 1973, page 65.
(56) Yu.I. Yermakov, B.N. Kuznetsov and V.L. Kuznetsov,
      Kinetika i Kataliz, 14, 1085 (1973).
(57) Yu.I. Yermakov and B.N. Kuznetsov, *Doklady Akademii
      Nauk, 207,* 644 (1972).
(58) Yu.I. Yermakov, B.N. Kuznetsov and Yu.A. Rindin, *Kine-
      tika i Kataliz, 14,* 1594 (1973).
(59) Yu.I. Yermakov and B.N. Kuznetsov, Reaction Kinetics
      and Catalysis Letters, Budapest, *1,* 87 (1974).
(60) O.N. Kimkhai, B.N. Kuznetsov, G.K. Boreskov and Yu.I.
      Yermakov, *Doklady Akademii Nauk, 214,* 146 (1974).

Ethylene Polymerization with the Catalysts of One and Two Component Systems Based on Titanium Trichloride Complex

S. FUJI

Research Center
Mitsui Petrochemical Industries, Ltd.
Yamaguchi, Japan

SYNOPSIS

The active species of three different catalyst systems were investigated to determine their chemical structures, valence of titanium involved, number of active centers, rate of propagation per active center, and properties of the polymer obtained. The catalyst systems studied were: one consisting of titanium trichloride complex, obtained by reducing titanium tetrachloride with diethyl aluminium chloride; a second having titanium tetrachloride added to the first as a cocatalyst, and a third having diethyl aluminium chloride added to the first as a cocatalyst. While the three systems gave rise to marked differences in the resulting rate of propagation, numbers of branches and double bonds in the polymers produced, and degrees of polymerization, it was considered that their active species shared similar chemical structure, and that the titanium present in them was trivalent. The principal difference among these catalysts was postulated to be the physical state around an active Ti(III) fixed on the solid catalyst.

1) INTRODUCTION

Considerable light has been shed on the chemical structure of the active species and on the valence of titanium involved in ethylene polymerization using Ziegler catalysts, particularly those that are soluble, and which are believed to form relatively uniform active species(1,2). But for heterogeneous Ziegler catalyst, available information on the chemical structure of active species has not progressed beyond inferences drawn from what is known about the homogeneous system, presumably due to the system being extremely complicated(3-8). Moreover, the trivalent titanium though generally assumed to be the active species in the heterogeneous system, remains to be proven.

A variety of methods has been proposed for determining the number of active centers (C*)(9-11), yet no comprehensive studies have been reported relating the active species and its number with the rate of propagation per active center (R_p*) and the properties of polyethylene obtained. Such knowledge is important to commercial production of polyethylene by Ziegler catalysis which requires control at will of the properties of the polymer.

The present report is an attempt to correlate the active species with the properties of the polyethylene formed. Three catalyst systems were investigated: a system that consists of a solid $TiCl_3$-complex obtained by reducing $TiCl_4$ with Et_2AlCl (DEA); a second system consisting of $TiCl_4$ added to the first as cocatalyst; and a third being a system that combines the first with DEA added as a cocatalyst. They are designated as catalyst systems A, B, and C, respectively. These catalyst systems are considered to be an excellent subject to afford a key to the problem, because the $TiCl_3$-complex alone is catalytically active in ethylene polymerization(12) and, with the cocatalysts added to it, not only its activity but also the properties of polyethylene obtained can be modified.

First, the alkyl groups present in these $TiCl_3$-complexes were examined for their thermal stability; the amount of thermally unstable alkyl groups (R_d; 'd' for 'decompose') and of those that are thermally stable (R_s; 's' for 'stable') were respectively determined to test their correlation with the catalytic activity. These catalysts were then subjected to ESR analysis to look into the valence of the titanium constituting the active species. This was followed by determination of the number of active centers (C*), and the rate of propagation (R_p*) per active center using $C^{14}O$ as the tracer (13). From all these results these factors which govern the R_p* in these catalyst systems and the properties of the polyethylene produced are elucidated.

2) *EXPERIMENTAL*

2.1) Materials

Commercially available $TiCl_4$, Et_2AlCl, and $EtAlCl_2$ were used as received. Commercially available ethylene of purity

over 99.9% was used as received. Kerosene used had b.p. 180°
∿200°, and a bromine number 0.01.

$C^{14}O$ was obtained from formic acid-C^{14} (The Radiochemical
Centre, Amersham) having specific radioactivity of 59.2 mCi/
mmole. Cold formic acid was added to formic acid-C^{14}, and
diluted to 53.4 μCi/mmole. The mixture was thermally decom-
posed at 60° with concentrated sulfuric acid to obtain $C^{14}O$.
It was used for quenching polymerization after drying it with
anhydrous calcium oxide.

2.2) Preparation of TiCl₃-complex catalyst

To a solution of $TiCl_4$ in hexane (1 mole/l) maintained
at 15° under vigorous stirring was added dropwise for 2 hr.
alkyl aluminium in the required quantity. After raising the
temperature to 25° in 10 min, the reaction was allowed to
continue for additional 24 hr. Solid TiCl₃-complex obtained
was washed with hexane and dried. All the procedures were
performed under nitrogen.

2.3) Analysis of TiCl₃-complex catalyst

Ti, Al and Cl were determined by chemical analysis.
Alkyl groups were analysed as follows, with all the procedures performed under nitrogen:

2.3.1) Thermal Decomposition (Determination of R_d).—In a
50 ml flask connected to a gas burette, 5 g of TiCl₃-complex
was precisely weighed, and heated at 140° for 10 min. Twice
the gas quantity actually evolved was taken to evaluate R_d,
since by disproportionation reaction, one half of the alkyl
radicals generated in the thermal decomposition would turn
into an inactive saturated hydrocarbon and the other half
into an unsaturated hydrocarbon, the latter is polymerized
in the presence of the Ti-catalyst.

2.3.2) Hydrolysis (Determination of R_s and R_t; 't' denoting
'total').—For determining R_t, 1 g of TiCl₃-complex was precisely weighed, whereas for determining R_s, 1 g of the solid
products of thermal decomposition of TiCl₃-complex was precisely weighed. The same apparatus as described above was
used. Each was hydrolysed with 10 ml of distilled water
added with a syringe through the rubber stopcap of the flask
maintained at 25°. R_t and R_s were calculated on the assumption that no isomerization reaction occurred in the course
of hydrolysis.

2.3.3) ESR Measurement.—In a quartz tube 5 mm in diameter was precisely weighed 1 g of TiCl$_3$-complex under nitrogen. A Nippon Denshi JES–ME–360 spectrometer was used with magnetic modulation at 100 kHz, and 10 gauss. DPPH was used as an internal standard to determine the g values and spin concentrations.

2.4) Polymerization and characterization

In a 500 ml four-necked separable flask were placed, at 60° under nitrogen, 250 ml of kerosene and the following amounts of catalyst components: TiCl$_3$-complex taken in a quantity of 5 mmole as Ti(III), (catalyst system A); 2.5 mmole of TiCl$_4$ added to the first, (catalyst system B); and 2.5 mmole of DEA added to 2.5 mmole of TiCl$_3$-complex, (catalyst system C). After immediately replacing N$_2$ with ethylene in the apparatus, polymerization was performed with vigorous stirring under atmospheric pressure at 60° for 1 hr. Rates of polymerization were calculated from quantities of ethylene absorbed during the reaction. Intrinsic viscosity was measured in decalin at 135°.

2.5) Determination of number of active centers (C*) and Rate of Propagation per active center (R$_p$*)

The C^{14}O quenching technique was chosen to determine C* because it obviates the necessity to correct for those polymer-aluminium linkage generated by chain transfer to alkyl aluminium during the polymerization, and because it is believed to be the most reliable method yet achieved(13). The measurement was performed as follows:
In a 100 ml flask was placed 50 ml of kerosene with 1 mmole of TiCl$_3$-complex (catalyst system A). For catalyst systems B and C, the cocatalysts were added to TiCl$_3$-complex in identical molar ratio to that given in Section 2.4. Polymerization of ethylene followed at 60° for 15 min, during that time the quantities of ethylene absorbed were measured throughout. After the purging of unreacted ethylene, growing chain ends were quenched with 15 ml of C^{14}O having specific radioactivity of 53.4 µCi/mmole.
The number of C^{14}O-labelled polyethylene molecules (C*) was calculated as follows:

$$C^* = \frac{R \cdot Y}{T \cdot r} \text{ (mole/mole Ti based on TiCl}_3\text{-complex)} \qquad [1]$$

where R=Specific radioactivity of polymer (µCi/g); Y=Polymer yield (g); T=Quantities of TiCl$_3$-complex used (mmole); and r=specific radioactivity of $C^{14}O$ (μCi/mmole), and R_p^* values were derived as the following:

$$R_p^* = R_p/C^* \qquad [2]$$

where R_p is the rate of polymerization at the point of quenching (mole ethylene/sec atm·mole·Ti of $TiCl_3$-complex).

3) RESULTS

3.1) Composition of $TiCl_3$-complex

Table 1 shows the composition of $TiCl_3$-complex catalyst as molar ratio to Ti and as related to their preparative conditions.

The table indicates that $TiCl_3$-complex contains thermally unstable alkyl groups (R_d) as well as those that are thermally stable, but susceptible to hydrolysis (R_s). It is assumed that R_d in this instance corresponds to the alkyl group in the thermally unstable R-Ti linkage(14), while R_s is the alkyl group in the Al-R linkage and/or in the Ti---^{R}---$_{Al}$ linkage. Although the C_2H_5-Ti is supposed to be very unstable, it may possibly exist stably where embedded in the $TiCl_3$ crystals in small quantity. This is believed to be the case in the $TiCl_3$-complex(15).

It is seen that increasing the DEA/$TiCl_4$ molar ratio in preparation increases the R_s content in $TiCl_3$-complex and decreases the contents of R_d, Al and Cl. Table 1 also shows, for a typical catalyst, the gas composition from which R_d and R_s were calculated, and it is seen that the major constituent is ethane. Comparing the quantities of Cl actually measured with those theoretically derived by assuming the Ti in $TiCl_3$-complex to be trivalent and the alkyl group to be attached on Al, their difference, ΔCl, is close to zero within experimental errors. It is consequently deduced that the overall stoichiometrical valence of Ti in the $TiCl_3$-complex is 3.

3.2) ESR spectra of $TiCl_3$-complex catalyst

The next problem is whether the valence of the Ti involved in the actual active species is likewise 3. It is known that trivalent Ti with an alkyl group as a ligand has low symmetry and can be observed in ESR(16). The ESR spectra were taken of the catalysts, each representative of the systems A, B, and C. Typical charts are shown in Fig. 1. For the catalyst system A, two signals are observed, one at g value of 1.903, and the other at a g value of 1.950. For B, signals are observed at the same positions but with dif-

TABLE 1.

Preparative conditions and composition of $TiCl_3$-complex catalyst.

Catalyst Number	Preparative Conditions Alkyl Al	Alkyl Al/Ti	Al	Cl	Molar Ratio/Ti $\hat{R}_t$	R_s	R_d	ΔCl^*
1	Et_2AlCl	0.4	0.464	4.24	0.119	0.079	0.082	+0.02
2	"	0.6	0.471	4.22	0.191	0.161	0.046	+0.02
3	"	0.8	0.371	3.77	0.238	0.206	0.036	-0.10
4	"	1.1	0.278	3.29	0.294	0.256	0.030	-0.26
5	"	1.5	0.280	3.53	0.295	0.274**	0.036***	0.00
6	$EtAlCl_2$	1.2	1.05	5.38	0.213	0.172	0.042	-0.12
7	"	2.2	1.13	5.98	0.313	0.267	0.016	-0.03

*ΔCl = Theoretical Cl - observed Cl,

Theoretical Cl was calculated from the formula: $TiCl_3 \cdot (AlCl_3 - nR^t_n)_x$.

Composition of evolved gas:

** Ethane 89.02, Propane 2.27, i-Butane 3.01, n-Butane 4.57, trans-Butane 1.13

*** Ethane 84.90, Propane 3.17, i-Butane 2.72, n-Butane 7.41, trans-Butane 1.80

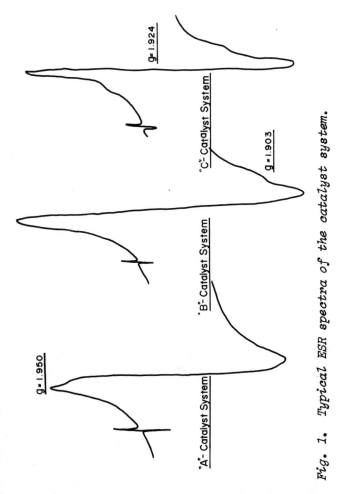

Fig. 1. Typical ESR spectra of the catalyst system.

ferent intensities. For C, an additional signal besides these two is observed, at a g value of 1.924.

Total spin concentrations (mole % based on Ti(III) of TiCl$_3$-complex) at 60° for the typical catalyst systems were determined and are tabulated in Table 2.

141

TABLE 2.

Spin concentration, number of active centers (C^*) and rate of propagation per active center (R_p^*).

Catalyst System		Spin Concentration	C^*	R_p^*
TiCl$_3$-complex Number	Co-catalyst	Mole % based on Ti of TiCl$_3$-complex	"	Mole of Ethylene / Mole of Ti·sec·atm
1	Et$_2$AlCl	0.070	0.17	34
3	none	0.014	0.019	124
4	none	0.047	0.027	92
	TiCl$_4$	0.075	0.060	216
5	Et$_2$AlCl	0.067	0.42	26
	none	0.042	0.028	90
	TiCl$_4$	0.061	0.052	272
	Et$_2$AlCl	0.047	0.35	40

3.3) Ethylene polymerization

Table 3 gives the results of ethylene polymerization, where the catalystic activity is expressed in terms of maximum rate of polymerization, as well as the characteristics of the polyethylene obtained.

No polymerization was observed for TiCl$_3$-complex obtained by reduction with EtAlCl$_2$ (No. 6 & 7). The catalytic activity of TiCl$_3$-complex obtained by reduction with DEA increased with the increase of DEA/TiCl$_4$ molar ratio in the catalyst preparation.

TABLE 3.

Results of ethylene polymerization.

| Catalyst Systems | | | Polymerization | | $[\eta]$ | Polymer Structure | | | |
TiCl₃-Complex Number	Cocatalyst Material	Amount (mmol)	Maximum Rate*	Yield (g)	(dl/g)	CH₃	IR (Number/1000C) tr-vinylene	vinyl	vinylidene
1	none	-	1.7	4.9	3.72	1.9	-	0.14	-
	TiCl₄	2.5	5.5	11.8	3.03	3.1	0.08	0.12	0.04
	Et₂AlCl	2.5	9.2	24.7**	16.5	1	-	-	-
2	none	-	4.5	24.7	5.36	1.7	-	0.19	0.07
	TiCl₄	2.5	8.8	32.7	3.66	1.9	0.06	0.13	0.09
	Et₂AlCl	2.5	14.0	29.8**	15.5	1	-	-	-
3	none	-	5.3	24.1	10.4	1	-	0.07	0.06
	TiCl₄	2.5	12.4	42.1	2.5	1	0.10	0.17	0.11
	Et₂AlCl	2.5	19.2	42.2**	13.1	1	-	-	-
4	none	-	6.6	29.1	12.2	1	-	-	-
	TiCl₄	2.5	14.5	47.5	2.53	2.1	0.13	0.21	0.13
	Et₂AlCl	2.5	23.0	60.4**	16.1	1	-	-	-
5	none	-	8.2	46.4	12.5	1	-	-	-
	TiCl₄	2.5	16.0	78.9	2.42	2.0	0.08	0.20	0.12
	Et₂AlCl	2.5	24.3	64.4**	13.4	1	-	-	-
6	none	-	0	0	-				
	TiCl₄	2.5	0	0	-				
7	none	-	0	0	-				
	TiCl₄	2.5	0	0	-				

(Polymerization conditions : Kerosene 250 ml, at 60°C, 1 hr)

* ℓ ethylene/hr mmole Ti of TiCl₃-complex.

** Ti³⁺= 2.5 mmol; for others, Ti³⁺ = 5 mmol.

143

Upon comparing the three different catalyst systems based on the identical TiCl$_3$-complex, the intrinsic viscosities of the formed polyethylene are in the order: $C>A>B$, while the number of branches and double bonds tends to be in the reversed order: $C<A<B$.

In order to examine the role of the alkyl groups, R_d and R_s, in the ethylene polymerization, a relationship between the amounts of the alkyl groups and the catalytic activity was sought. Fig. 2 indicates an inverse dependence between the catalytic activity of the system A and the quantity of R_d. Fig. 3 shows that the catalytic reactivities of A, B and C are proportional to the quantity of R_s.

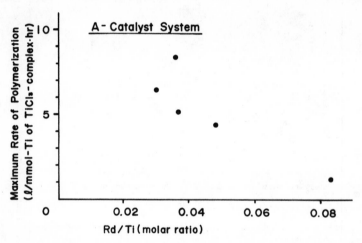

Fig. 2. Relationship between maximum rate of polymerization vs R_d.

The system TiCl$_3$-complex + TiCl$_4$ + EtAlCl$_2$ is known as Ziegler-Martin catalyst(17). It is noted that where DEA is added as a cocatalyst to α-TiCl$_3$(7) and TiCl$_3$-complex(18), adsorption equilibrium of DEA onto TiCl$_3$ is responsible for the ensuing active species. However, no studies have yet been reported of the catalyst system B wherein no alkyl aluminium is used as a cocatalyst. Therefore, the effect of TiCl$_4$ to activate TiCl$_3$-complex in case of the system B was investigated, especially from the aspect of whether the reaction of TiCl$_4$ with TiCl$_3$-complex is an adsorption equilibrium.

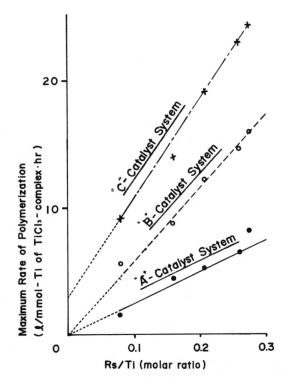

Fig. 3. Relationships between maximum rate of polymerization vs R_s.

Comparison was made among the catalytic activities of the following three catalysts: $TiCl_3$-complex (A); $TiCl_4$ added to A, (B); and the $TiCl_3$-complex having been treated as follows: the above second catalyst was prepared and aged under nitrogen at the same temperature as that of polymerization, followed by washing with kerosene. The results indicated that the catalytic activity for the last one was lower than that of the second and was of approximately the same activity as that of the first. It was also found that most of the Ti in the supernatant was in the form of $TiCl_4$, and soluble trivalent Ti was not detected. These are indications of the fact that $TiCl_4$ adsorbed reversibly onto the $TiCl_3$-complex to form active centers.

3.4) Number of active centers (C^*) and rate of propagation (R_p^*)

The results in Table 2 indicate that C^* tends to increase

slightly as the molar ratio of Al/Ti is increased in the preparation of $TiCl_3$-complex, as long as the kind of cocatalyst
used is retained identical; that R_p^* remains constant within
experimental errors; and that, the values of C^* and R_p^* undergo drastic changes with different kinds of cocatalyst used.
The values of C^* increase in the order of the catalyst
systems: $A<B<C$, viz., A (0.018~0.028 mole % based on Ti^{3+} of
$TiCl_3$-complex), B (0.052~0.060 mole %), and C (0.17~0.42 mole
%). R_p^* increases as the catalyst system is changed, in the
order of $C<A<B$, viz., C (26~40 mole ethylene/mole C^* sec atm;
the same unit applying in what follows), A (90~124) and B
(216~272).

4) DISCUSSION

4.1) Chemical structure of active species

4.1.1) $TiCl_3$-Complex Catalyst (Catalyst System A).—Assuming
that the active species of this catalyst system consists of
simple R-Ti bonding (19-21), as has been suggested by Cossee,
Beerman et al., R_d would be expected to correlate positively
with the catalytic activity. However, as may be seen from
Fig. 2, no such result was obtained. Therefore, the Cossee's
model does not explain the chemical structure of the active
species of this catalyst system. On the other hand, Fig. 3
shows that the catalyst system A has its catalytic activity
directly proportional to R_s. Consequently, it was postulated
that R_d has nothing whatever to do with the active species,
and that the active species of this catalyst lies in R_s.

As previously mentioned, R_s consists of either or both
R-Al bonding and $Ti\overset{R}{\diagdown}Al$ bonding. Since the possibility
of R-Al bonding alone constitutes the active species of
ethylene polymerization can be ruled out(19), the bridged
structure of $Ti\overset{R}{\diagdown}Al$, or Natta's model(7,22-25), is considered to constitute the active species. The multiplicity of
the ESR signals (Fig. 1) indicates the existence of different
kinds of chemical species in the catalyst A. The following
chemical structures of the active species may probably be
considered:

(Ia)   (Ib)

$$- \begin{array}{c} Cl \\ | \\ Ti \\ | \\ Cl \end{array} - Et - Cl - \quad - Al - Cl -$$

(with vacant orbital box above Ti)

(*II*)

The structures of type *I* may only be derived from dialkyl Al compound, while *II* may be from both di- and monoalkyl Al compounds. The TiCl$_3$-complex obtained by reduction with mono-alkyl Al, i.e., EtAlCl$_2$, had no catalytic activity (Table 3, No. 6 & 7), and the possibility of the structure *II*, consequently, could be excluded.

4.1.2) TiCl$_3$-complex + TiCl$_4$ system (Catalyst System *B*).—It is seen from Fig. 2 that the catalyst system *B* has catalytic activity twice as high as that of *A*, and that its catalytic activity is proportional to R$_S$. This can be explained by considering that a portion of R$_S$ reacts with the cocatalyst TiCl$_4$ to form new active centers, besides the type *I* inherent to TiCl$_3$-complex. Since it is unlikely that TiCl$_4$ reacts with R$_S$ in the $_{Ti}$$\diagdown^{R}\diagup_{Al}$ bond, it possibly reacts with that in Al-Et to form the active species. Upon considering the fact that TiCl$_4$ is adsorbed onto the TiCl$_3$-complex reversibly in the formation of active centers, the following reversible reaction could be written:

$$\begin{array}{c} Et \\ \diagup \quad \diagdown \\ Al \\ \diagup \quad \diagdown \\ Cl \qquad Et \end{array} + TiCl_4 \rightleftharpoons \begin{array}{c} Cl \quad Cl \\ \diagdown \; | \; \diagup \\ Ti \\ | \\ Cl \\ - Cl - - Et - \\ Al \\ Cl \diagdown \quad \diagup Et \end{array}$$

(*III*)

where ▢ represents a vacant orbital.

The aluminium alkyls on the surface of TiCl$_3$-complex with which TiCl$_4$ reacts, may have either one or two ethyl groups attached to them as shown in *Ia/Ib* and *II*. The structure involving one alkyl group could be, again, excluded for the same reasons as given for the catalyst system *A*.

4.1.3) TiCl$_3$-Complex + DEA System (Catalyst System *C*).—For the catalyst system *C*, the effects of the molar ratio of Al/ Ti in polymerization on the catalytic activity as well as on properties of the polyethylene formed are very similar to those observed for α-TiCl$_3$ + DEA system(18). Considering, in

147

addition, the fact that where DEA is added as a cocatalyst to
α-TiCl$_3$(7) and TiCl$_3$-complex(18), adsorption equilibrium of
DEA onto TiCl$_3$ is responsible for the ensuing active species,
the chemical structure of the additional species beside I may
conceivably be analogous to the following structure IV, that
has been suggested for the system consisting of α-TiCl$_3$ +
Et$_3$Al(4,7).

$$(IV)$$

Fig. 2 shows that the catalytic activity of this
catalyst system is positively correlated with the R$_s$. The
interpretation for this would be that the increase of Al/Ti
molar ratio in the preparation of the catalyst results in the
increased number of the sites of adsorption of DEA, probably
in the crystalline region of TiCl$_3$, as well as in the increased quantity of R$_s$.

4.2) Valence of Ti in active species

The spin concentrations for the catalyst systems A and
B, as observed by ESR (Table 2), are fairly close to C* determined in the course of ethylene polymerization. Hence, the
Ti constituting the active species in the ethylene polymerization with these catalyst systems is very possibly trivalent.
Fig. 4 shows that, for both A and B, the rate of polymerization starts out high from the very outset of polymerization,
gradually declining as the polymerization time elapses. This
indicates that the active centers in these two catalyst systems were already present in the catalyst prior to polymerization.
With the catalyst system C, on the other hand, its ESR
spin concentration is less than C*. According to the studies
by Keii(26), with the system consisting of α-TiCl$_3$ + DEA,
whose active species are generally believed to consist of a
trivalent titanium(7,27), the spin concentration was found
to increase with the passage of polymerization time. This
may be taken to suggest that the ESR spin concentration prior
to polymerization tends to be lower than C* measured during
polymerization. In the author's experiments, too, the rate

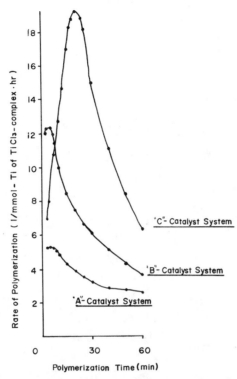

Fig. 4. Rate of polymerization and polymerization time. (#3 TiCl₃-complex was used).

of polymerization with C was actually noted to increase gradually with the passage of polymerization time (Fig. 4). Consequently, the titanium in the active species of C is reasonably considered to be trivalent.

4.3) Principal differences among active species

The catalyst systems and their active species are shown in comparison with molecular structure (branches, double bonds) and intrinsic viscosity of the polymer and R_p^* in Table 4, where we find both R_p^* and the characteristics of polyethylene obtained differing widely among the catalyst systems.

Considering the elementary reactions involved in Ziegler-Natta polymerization (27–32), the increase of the quantity in branches and double bonds together with the decrease of $\overline{M}$ indicates the increased occurrence of the transfer reaction

TABLE 4.

Characteristics of catalyst systems.

Catalyst System	C	A	B
	(IV)	(I)	(III)
R_p*	slow	medium	fast
$[\eta]$	high	medium	low
Branches, Double Bonds	few	medium	many

150

involving the β-hydrogen of the growing chain end(33–35), in other words, the decrease in bonding force of the Ti-R linkage at a growing chain end. The difference in the rate of propagation ($R_p{}^*$) could also be interpreted on the basis of the bonding strength of the Ti-R linkage.

The question now arises as to what it is that differentiates the behavior of these catalyst systems. Upon inspecting the chemical structures and the valence of Ti of each active center discussed so far, no significant difference can be found. However, comparing the physical state surrounding the active Ti(III) of the catalyst systems A, B and C, large difference may be noticed. More specifically, it is presumed that, in the catalyst system C, the formed active species is firmly embedded in the crystal lattice of TiCl₃. In the catalyst system A, while its active species is located in TiCl₃-complex, it is presumed that its peripheral crystal lattice is disturbed because the bulky ethyl groups have replaced Cl which served as ligands to Ti. This would make its active species less tightly bonded in the solid catalyst than is the case with the catalyst system C. In the catalyst system B, its active species is presumed to be only weakly chemically adsorbed onto the surface of TiCl₃-complex.

In view of the differences both in the chemical structure of the polymer and in the nature of the catalyst systems, one feels compelled to conclude that the extent of firmness of an active Ti(III) fixed on TiCl₃-complex affects the bonding strength of the R-Ti linkage, which in turn affects $R_p{}^*$ and the characteristics of the polymers formed.

Schindler(33) has reported that the transfer reaction involving the β-hydrogen of the growing chain end does occur with the catalyst system TiCl₄ + DEA, but does not with the system TiCl₃ + DEA. The author believes that here, too, one should duly reckon with the fact that the active species in the system TiCl₄ + DEA is nearly soluble(34), and that the active species in the system TiCl₃ + DEA is firmly bonded in the solid catalyst.

ACKNOWLEDGMENT

The author is indebted to Mr. A. Otsuka, of Otsuka Seiyaku Co., Ltd., for his generosity in making his RI equipment available for use in the present study. Valuable aids were provided by many of the author's colleagues, among whom are to be mentioned, Dr. A. Moriuchi who made the ESR measurements, and the members of the Analytical Division, Iwakuni-Otake Works, Mitsui Petrochemical Industries, Ltd., who made the IR and chemical analysis of the catalysts and measurements of intrinsic viscosities of polyethylene. Last but not least, much inspiration was received from the thoughtful discussion afforded by Dr. A. Takeda, Director, Research Center, Mitsui Petrochemical Industries, Ltd.

LITERATURE CITED

(1) G. Natta, P. Pino, G. Mazzanti, U. Guannini, E. Mantica, and M. Peraldo, *J. Polymer Sci. 26*, 120 (1957).
(2) G. Henrici-Olivé, and S. Olivé, *Angew. Chem. 79*, 764 (1967).
(3) H. Schnecko, *Makromol. Chem. 84*, 156 (1965).
(4) L.A.M. Rodriguez, and H.M. Van Loog, *J. Polymer Sci. A-1, 4*, 1971 (1966).
(5) L.A.M. Rodriguez, and H.M. Van Loog, *Polymer Reviews, 1*, 259 (1965).
(6) L.A.M. Rodriguez, and H.M. Van Loog, *J. Polymer Sci. A-1, 4*, 1951 (1960).
(7) L.M. Lanovskaya, N.V. Makletsova, A.R. Gantmakher, and S.S. Medvedev, *Vysokomol. Soed. 7*, 747 (1965).
(8) L. Kollar, A. Simon, and J. Osvath, *J. Polymer Sci. A-1*, 919 (1968).
(9) B. Bier, W. Hoffmann, G. Lehmann, and G. Seydel, *Makromol. Chem. 58*, 1 (1962).
(10) W. Cooper, D.E. Eaves, G.D.T. Owen, and G. Vaugham, *J. Polymer Sci. C4*, 211 (1964).
(11) H.W. Coover, Jr., J.E. Guillet, R.L. Combs, and F.B. Joyner, *J. Polymer Sci. A-1, 4*, 2583 (1966).
(12) M.L. Cooper, and J.B. Rose, *J. Chem. Soc.*, 795 (1959).
(13) Yu.I. Ermakov, V.A. Zakharov, and G.D. Bukatov, *Catalysis 1*, 399 (1973), Proceeding of the Fifth International Congress on Catalysis, Miami Beach, Fla., 20-26 August (1972).
(14) C. Beerman, *Angew. Chem. 71*, 618 (1959).
(15) A. Schindler, *Die Makromol. Chem. 102*, 263 (1967).
(16) B.N. Figgis, Introduction to Ligand Fields (Interscience Publishers, Division of John Wiley and Sons, New York).
(17) T.P. Wilson, and G.F. Hurley, *J. Polymer Sci. C*, 281 (1963).
(18) S. Wada (Mitsui Petrochemical Industries), unpublished data.
(19) P. Cossee, *J. Catalysis 3*, 80 (1964).
(20) C. Beerman, and H. Bestian, *Angew. Chem. 71*, 618 (1959).
(21) G. Van Heerden, *J. Polymer Sci. 34*, 46 (1959).
(22) G. Natta, *J. Polymer Sci. 34*, 21 (1959).
(23) F. Patat, and H. Sinn, *Angew. Chem. 70*, 496 (1958).
(24) H. Velzman, *J. Polymer Sci. 32*, 457 (1958).

(25) J. Furukawa, and T. Tsuruta, *J. Polymer Sci.* *36*, 275 (1959).

(26) Y. Ono, and T. Keii, *J. Polymer Sci.* *A-1*, *4*, 2441 (1966).

(27) E.J. Arlman, *J. of Catalysis 5*, 178 (1966).

(28) G. Natta and G. Mazzanti, *Tetrahedron 8*, 86 (1960).

(29) A. Zambelli, I. Pasquon, R. Signorini, and G. Natta, *Makromol. Chem. 112*, 160 (1968).

(30) G. Natta, I. Pasquon, A. Zambelli, and G. Gatti, *J. Polymer Sci. 51*, 387 (1961).

(31) I. Pasquon, *Makromol. Chem. 3*, 465 (1967).

(32) L. Reich, and S.S. Stivala, *J. Polymer Sci. A1*, 203 (1963).

(33) A. Schindler, *Polymer Letters 4*, 193 (1966).

(34) J. Varadi, I. Czajlik, A. Baan, N.M. Chirkov, and V.I. Tsvetkova, *J. Polymer Sci. C*, 2069 (1967).

(35) H. Bestian and K. Clauss, *Angew. Chem. 75*, 1068 (1963).

A Kinetic Model for Heterogeneous Ziegler-Natta Polymerization

PETER J.T. TAIT

Department of Chemistry
U.M.I.S.T., Manchester, U.K.

Ziegler-Natta polymerization reactions have been the subject of extensive kinetic studies and a number of attempts have been made to account for their kinetic behaviour. All of these attempts have necessitated the adoption of assumptions concerning the nature of many of the fundamental processes, both chemical and physical, which are involved in these fascinating yet complex polymerization systems. Eirich and Mark(1) were the first to point out that, since most Ziegler-Natta catalyst systems were heterogeneous in nature, it was most likely that adsorption or complex formation processes were involved in such stereospecific polymerization reactions. These concepts have been used by a number of workers and in particular by Saltman(2) who proposed that monomer and metal alkyl molecules were reversibly adsorbed onto the crystal surface of the transition metal halide, and that chain propagation occurred between the adsorbed metal alkyl and monomer, either adsorbed or in solution. At high metal alkyl concentrations the derived expression for the rate of propagation, assuming that propagation is with unadsorbed monomer, is given by:

$$R_p = k_p[M][S] \qquad [1]$$

where $[S]$ is the concentration of surface sites, $[M]$ is the monomer concentration and k_p is the rate constant for chain propagation. This equation is similar to expressions derived by Kern *et al.*(3) and Keii *et al.*(4), for the polymerization of propylene by $TiCl_3/AlEt_3$.

Eq. 1 does not, however, have general application for, although it is consistent with the first order dependence of the overall rate of polymerization on monomer concentration, and consistent with the independence of rate at high metal alkyl concentrations(3-5), it does not explain the maxima in rate observed by some workers as the metal alkyl concentration is increased. A maxima in rate has been observed in systems(6,7-9) where it cannot be attributed to deactivation by reduction of the active species as has been suggested by Kern *et al.*(3), although this latter explanation is probably important for catalysts based on higher valence state transition metal halides, e.g., $TiCl_4$.

Vesely *et al.*(10) on the other hand have proposed a kinetic scheme where propagation is considered to occur between adsorbed metal alkyl and adsorbed rather than unadsorbed monomer. Under these circumstances the rate equation is of the form:

$$R_p = k_p \theta_M \theta_A S \qquad\qquad [2]$$

where θ_M and θ_A are the fractions of surface S covered by adsorbed monomer and adsorbed metal alkyl respectively. This equation correctly predicts a fall off in rate at high metal alkyl concentrations, but the model adopted requires that the active center concentration, i.e., concentration of growing chains, should be proportional to the fraction of surface covered by adsorbed monomeric metal alkyl. This has been shown to be not the case for some systems(11).

The above reaction schemes are based on the assumption that the active centers are adsorbed metal alkyl species. The active centers, however, may be alkylated transition metal entities, as proposed by Cossee(12). Otto and Parravano (6) have proposed a polymerization scheme in which chain propagation occurs between adsorbed monomer and an alkylated transition metal entity, the concentration of which is proportional to the fraction of the surface covered by adsorbed metal alkyl. However, their derivations do not appear to have general application outside their own kinetic results.

Although these approaches can successfully explain some of the observed kinetic features of Ziegler-Natta polymerization reactions, they are all somewhat limited to particular systems, or to particular concentration ranges. A kinetic model will now be presented which adequately describes the extensive kinetic, molecular weight and active center data obtained for the system $VCl_3/AlR_3/4$-methylpentene-1, and which appears perhaps to have a fairly general application to many other Ziegler-Natta polymerization systems(11). It will be seen that in the simple kinetic model which is proposed that equilibrium adsorption processes are envisaged as playing an important role in determining the kinetic course of these polymerization processes.

1) PROPOSED KINETIC SCHEME

The active centers for the polymerization of 4-methyl-pentene-1(4-MP-1) by the catalyst system $VCl_3/Al(i\text{-}Bu)_3$ are considered to be alkylated entities, such as VCl_2R, on the surface of the vanadium trichloride crystals, rather than metal alkyl molecules merely adsorbed onto the surface. Chain initiation is believed to arise from the following sequence of reactions:

$$VCl_3 + AlR_3 \longrightarrow VCl_2R \cdot AlR_2Cl \quad (fast) \qquad [3]$$

$$VCl_2R \cdot AlR_2Cl + AlR_3 \rightleftharpoons VCl_2R + Al_2R_5Cl \quad (slow) \qquad [4]$$

$$VCl_2R + M \longrightarrow VCl_2P \qquad [5]$$

VCl_2P is equivalent to C_0 and eq. 5 is the initiation process. Chain propagation is believed to occur between an active center and adsorbed monomer, both monomer and alkyl molecules being able to compete for complex formation with active centers:

$$VCl_2P + M \underset{}{\overset{K_M}{\rightleftharpoons}} VCl_2P \cdot M \qquad [6]$$

$$VCl_2P + A \underset{}{\overset{K_A}{\rightleftharpoons}} VCl_2P \cdot A \qquad [7]$$

The equilibrium nature of reactions 4, 6 and 7 is believed to play an important part in determining the kinetic course of these polymerization reactions.

The term adsorption, as used in this paper, relates to weak chemisorption of species onto the active centers on the catalyst surface, i.e., where chemical bonding of some form is involved. Thus only those species which are capable of specific interaction with the active sites are considered as adsorbed, any physisorption will be much weaker and is neglected. Hence monomer may be adsorbed through complex formation of a π-donor complex, e.g.,

$$PCl_2V \leftarrow \underset{\underset{R}{\overset{|}{CH}}}{\overset{CH_2}{||}}$$

Aluminum alkyl species are probably adsorbed through the formation of bridge structures such as:

The adsorption of monomer and metal alkyl onto the catalyst surface are now considered to be described by Langmuir-Hinshelwood isotherms of the forms:

$$\Theta_M = \frac{K_M [M]}{1 + K_M[M] + K_A[A]} \qquad [8]$$

and

$$\Theta_A = \frac{K_A [A]}{1 + K_M[M] + K_A[A]} \qquad [9]$$

where [M] and [A] are the equilibrium concentrations of monomer and metal alkyl, and K_M and K_A are the equilibrium constants for the respective adsorption equilibria. It should be noted that the K_M and K_A values quoted are calculated with respect to factors affected by adsorption at the active center. Consequently, the values so obtained may not be the same as those relating to adsorption on unreacted VCl_3 crystal surfaces.

Under these circumstances and adopting the proposed model for the polymerization reaction the overall rate of polymerization will be given by:

$$R_p = k_p \Theta_M C_0 \qquad [10]$$

where k_p is the propagation rate constant with respect to adsorbed monomer, Θ_M is the fraction of the active surface covered by adsorbed monomer, and C_0 is the active center concentration. Hence:

$$R_p = \frac{k_p K_M[M] \ C_0}{1 + K [M] + K [A]} \qquad [11]$$

This model has been successfully used in an analysis of kinetic, metal-polymer bond(MPB), and molecular weight dependencies observed for the polymerization of 4-MP-1 by the catalyst system VCl_3/AlR_3 and by the modified system VCl_3/AlR_3 Et_3N.

The polymerization system $VCl_3/AlR_3/4$-MP-1 was selected as a model system for this study because the various components could be obtained in a very high degree of purity, and because this particular polymerization system is stable and not subject to undesirable side reactions. Reproducible results in the temperature range 30-45°C can be obtained which have the required degree of accuracy to enable the necessary analyses to be carried out successfully.

2) EXPERIMENTAL PROCEDURE

Polymerizations were carried out in special dilatometers as have been described elsewhere(13). All rates could be reproduced to within ±4%.

A quench technique(14) using tritiated methanol was used to determine the metal-polymer bond concentration. A value of 3.20 ± 0.15 was used as a correction factor to take into account the kinetic isotope effect(10).

Molecular weights were determined in decalin at 135°C by means of viscometry using an Ubbelohde suspended level dilution viscometer. The relationship,

$$[\eta] = 1.94 \times 10^{-4} \; (M_n)^{0.81} \qquad [12]$$

as derived by Hoffman *et al.*(15), was used in the calculation of number-average molecular weights for this study.

3) OVERALL KINETIC FEATURES

The polymerization reaction is characterized by an initial 'settling' period of increasing rate which is followed by a much longer period during which the rate gradually decreases due to depletion in monomer concentration. Results for a typical polymerization are shown in Fig. 1. After the

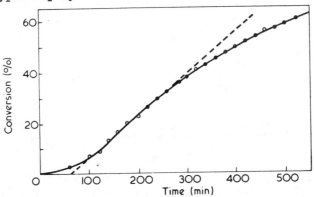

Fig. 1. Plot of conversion versus time. [4-MP-1]=2.00mole/l; [VCl₃]=18.5mmole/l; [Al(i-Bu)₃]=37.0mmole/l; temperature =30°C.

initial 'settling' period the polymerization rate, corrected for decrease in monomer concentration, remains constant up to at least 55% conversion. This behaviour is further confirmed by the results depicted in Fig. 2 which also show that the rate within a given polymerization reaction is first order with respect to monomer concentration. The 'settling' period is found by extrapolution of the linear portion of this first order plot to zero conversion as is illustrated in Fig. 2.

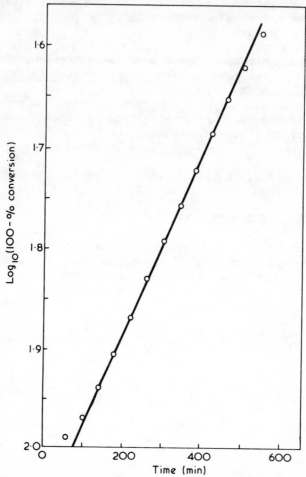

Fig. 2. Plot of $log_{10}(100-\%$ conversion) versus time. [4-MP-1]=2.0 mole/l; [VCl₃]=18.5 mmole/l; [Al(i-Bu)₃]=37.0mmole/l; temperature=30°C.

3.1) Dependence of Overall Rate on Catalyst Concentration

When the catalyst concentration is low, i.e., when the ratio $[4\text{-MP-1}]/[\text{Al}(i\text{-Bu})_3]$ is high, and when a constant monomer concentration and a constant tri-isobutylaluminium to vanadium trichloride molar ratio is employed, the steady state rate, R_p, i.e., the rate after the cessation of the 'settling' period, is found to vary linearly with the concentration of

TABLE 1.

Effect of catalyst concentration on rate of polymerization.[a]

$[VCl_3] \times 10^3$ (mole/l)	$\dfrac{[4\text{-MP-1}]}{[VCl_3]}$	Steady state rate x 10^3, (mole/l/min)	$\dfrac{Rate}{[VCl_3]}$
6.0	333	1.88	0.314
8.8	226	2.58	0.292
10.2	196	2.70	0.264
13.7	145	3.84	0.286
15.2	131	4.18	0.274
15.7	127	3.94	0.252
16.8	119	4.60	0.274
17.5	115	4.44	0.254
18.9	106	5.56	0.294
21.5	93	6.12	0.284
22.6	89	5.54	0.246

[a][4-Methylpentene-1]=2.0 mole/l; Al(i-Bu)$_3$/VCl$_3$ = 2.0; solvent = benzene; polymerization temperature = 30°C.

TABLE 2.

Effect of catalyst concentration on rate of polymerization.[a]

$[VCl_3] \times 10^3$ (mole/l)	$\dfrac{[4\text{-MP-1}]}{[VCl_3]}$	Steady state rate x 10^3 (mole/l/min)	$\dfrac{Rate}{[VCl_3]}$
3.9	519	2.34	0.606
4.8	414	3.38	0.698
7.4	272	4.72	0.642
7.4	270	4.80	0.648
7.5	268	4.70	0.632
8.7	229	5.64	0.646
11.2	178	7.04	0.628
14.7	137	8.90	0.608
18.4	109	12.30	0.668

[a][4-Methylpentene-1]=2.0 mole/l; Al(i-Bu)$_3$/VCl$_3$ = 2.0; solvent = benzene; polymerization temperature = 40°C.

vanadium trichloride. This dependence is shown in Tables 1 and 2. Logarithmic plots of the titanium trichloride concentrations *versus* steady state rates yield straight line plots of slope unity, indicating a first-order reaction with respect to titanium trichloride concentration. These results indicate, as is usual in Ziegler-Natta polymerization systems, that the number of active centers is directly proportional to the concentration of transition metal halide, i.e.,

$$C_0 \alpha [VCl_3] \qquad [13]$$

However, under conditions where the ratio $[4-MP-1]/[Al-(i-Bu)_3]$ is much lower, the steady state rate divided by the vanadium trichloride concentration decreases with increase in vanadium trichloride concentration, i.e., with decrease in the ratio $[4-MP-1]/[Al(i-Bu)_3]$. This is clearly evident from an examination of Table 3 and its significance will be discussed in the next section.

TABLE 3.

Effect of catalyst concentration on rate of polymerization.[a]

$[VCl_3] \times 10^3$ (mole/l)	$\dfrac{[4-MP-1]}{[VCl_3]}$	Steady state rate $\times 10^3$ (mole/l/min)	$\dfrac{Rate}{[VCl_3]}$
4.8	42	9.16	0.190
5.8	35	9.48	0.164
10.3	19	10.84	0.106
11.0	18	13.30	0.122
12.6	16	12.12	0.096
17.6	11	14.76	0.084
25.5	8	18.86	0.074

[a] $[4-Methylpentene-1] = 2.0$ mole/l; $Al(i-Bu)_3/VCl = 2.0$; solvent = benzene; polymerization temperature = 30°C.

3.2) Dependence of Rate on Tri-isobutylaluminium Concentration.

A distinct maximun in the rate of polymerization per unit vanadium trichloride concentration occurs at a molar ratio of catalyst components of approximately 2:1 when the concentrations of monomer and vanadium trichloride are maintained constant. This effect is shown for the catalyst system $VCl_3/Al(i-Bu)_3$ in Fig. 3. The actual number of moles of aluminium

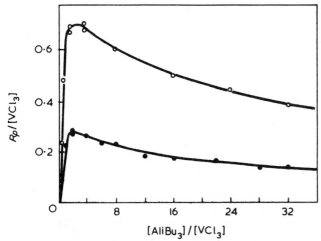

Fig. 3. *Rate of polymerization as a function of Al to V ratio at constant monomer and constant vanadium trichloride concentration.* $[4-MP-1]=2.0$ *mole/l;* $[VCl_3]\approx17.5$ *mmole/l at* $30°C$; $[VCl_3]\approx8.75$ *mmole/l at* $40°C$. ●, $30°C$; O, $40°C$.

alkyl which combines with the vanadium trichloride to form active sites does not necessarily correspond to the initial Al:V ratio, since the alkyl may assume several roles in the catalyst system. However, the decrease in rate above this maximum may be attributed to the excess aluminium alkyl competing with the monomer molecules for complex formation with active sites, and is in agreement with eqs. 6 and 7.

If however, the concentration of vanacium trichloride is varied whilst the amount of tri-isobutylaluminium and monomer are kept constant, the steady state rate per unit vanadium trichloride concentration increases to a limiting value as the concentration of vanadium trichloride is decreased, i.e., as the catalyst ratio Al:V is increased. This is shown in Fig. 4. Hence the fall off in rate described previously must be attributed to the increasing tri-isobutylaluminium concentration, and is not a consequence of the ratio of the catalyst components. Similar results were found for several other aluminium alkyls and are shown in Fig. 5.

If the active center for the polymerization is an alkylated entity on the surface of the vanadium trichloride, such as VCl_2R, then the increase in the rate of polymerization at constant vanadium trichloride concentration with increasing tri-isobutylaluminium concentration in the region up to Al:V=2:1 can readily be explained by considering the

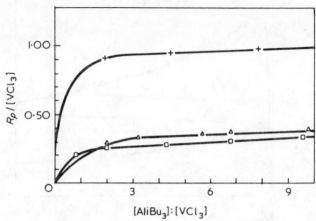

Fig. 4. Rate of polymerization as a function of Al to V ratio at constant monomer and constant aluminium tri-isobutyl concentration. [4-MP-1]=2.0mole/l; [Al(i-Bu)₃]=37.0mmole l. +, 50°C; □, 30°C; △, 30°C (aged catalyst).

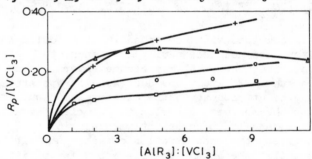

Fig. 5. Rate of polymerization as a function of Al:V ratio at constant monomer and constant aluminium alkyl concentration for different aluminium alkyls at 30°C. [4-MP-1]= 2.0 mole/l; [AlR₃]=37.0mmole/l. +, Al(n-Bu)₃; △, AlEt₃; O, Al(n-Hex)₃; □, Al(n-Dec)₃.

sequence of reactions 3, 4 and 5. Under these conditions it can be shown(16) that, the active center concentration C_0, is given by

$$C_0 \propto \frac{[VCl_3][AlR_3][M]}{[AlR_2Cl]}$$

[14]

Also since the concentration of vanadium trichloride, and consequently the concentration of aluminium chloroalkyl is constant:

$$C_o \propto \frac{[VCl_3][AlR_3][M]}{[AlR_2Cl]} \qquad [15]$$

Hence the initial increase in rate with increase in concentration of aluminium alkyl should be almost directly proportional to the concentration of aluminium alkyl since at these low concentrations, $\theta_A \propto [AlR_3]$. This is found to be the case as is shown in Fig. 6. C_o will have a maximum value when the whole of the potential active centers have become growing chains. Consequently at higher concentrations when the formation of active centers is almost complete, the steady state rate will begin to fall because of the lower values of θ_M. Hence above an Al:V ratio of 2:1 the rate of polymerization will be given by eqs. 10 and 11.

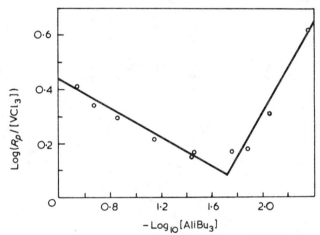

Fig. 6. Plot of log_{10} (rate) versus $log_{10}[Al(i-Bu)_3]$. [4-MP-1]=2.0 mole/l; [VCl$_3$]=8.7mmole/l; temperature=40°C.

3.3) Dependence of Rate on 4-Methylpentene-1 Concentration.

Linear plots of steady state rates of polymerization as a function of monomer concentration are obtained at 40°C, when the order of reaction with respect to monomer concentration, at constant catalyst concentration and constant Al:V ratio, can be shown to be 1.04±0.03. Such a plot is shown in Fig. 7 and is in agreement with the results shown in Fig. 2.

At 30°C however, although the points above 1.0 mole/l lie on a straight line passing through the origin, points below this concentration deviate considerably from a first-order dependence (Fig. 8).

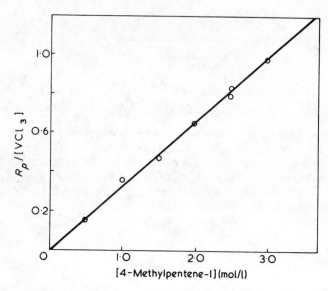

Fig. 7. *Variation in rate of polymerization with monomer concentration at 40°C.* $[VCl_3]=8.8mmole/l$; $[Al(i-Bu)_3]=17.6mmole/l.$

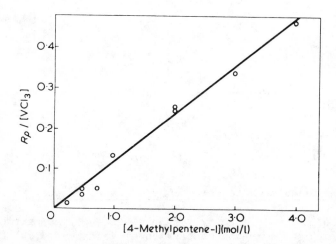

Fig. 8. *Variation in rate of polymerization with monomer concentration at 30°C* $[VCl_3]=17.8mmole/l$; $[Al(i-Bu)_3]=35.6$ *mmole/l.*

3.4) <u>Dependence of Rate on Different Aluminium Alkyl Compounds.</u>

The rate of polymerization is affected by the nature of the aluminium alkyl compound which is used to form the active catalyst. The results of using a series of aluminium alkyl compounds are summarized in Table 4.

TABLE 4.

Variation in rate with nature of aluminium alkyl.

Aluminium alkyl	$R_p/[VCl_3]$ (mole/l/min/mole VCl_3)	
	30°C	40°C
Trimethyl	0.288	0.286
Tri-isobutyl	0.280	0.660
Triethyl	0.253	0.700
Tri-n-butyl	0.221	---
Diethylchloride	1.169	0.110
Tri-n-hexyl	0.149	---
Tri-n-decyl	0.107	0.240

Catalyst systems based on $AlMe_3$, $AlEt_3$, $Al(n-C_{10}H_{21})_3$ and $Al(n-C_6H_{13})_3$ all exhibit the same type of kinetic behaviour as those using $Al(n-Bu)_3$, in that there is an initial 'settling' period which is followed by a period of steady rate. The kinetic behaviour, however, of the catalyst system $VCl_3/AlEt_2Cl$ is different in that the rate is found to increase rapidly to a maximum and then to decrease after only a short steady state period (Fig. 9).

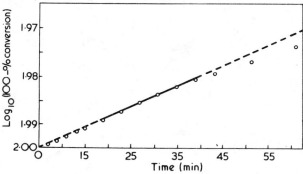

Fig. 9. Plot of $log_{10}(100-\%$ conversion) versus time using aluminium diethylchloride. [4-MP-1]=2.0 mole/l; [VCl_3]= 9.0 mmole/l; [$Al(i-Bu)_3$]=18.0 mmole/l; temperature=40°C.

4) *ACTIVE CENTER DETERMINATION.*

The MPB concentration increases continuously with polymerization time and with % conversion as is shown in Fig. 10 and 11. It has been shown previously that the rate of poly-

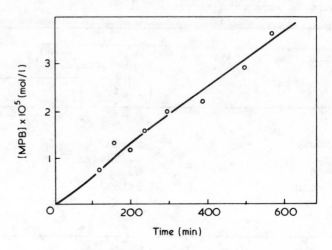

Fig. 10. Variation of MPB concentration with time. [4-MP-1]=2.0 mole/l; [VCl₃]=18.5mmole/l; [Al(i-Bu)₃]=37.0mmole/l; temperature=30°C.

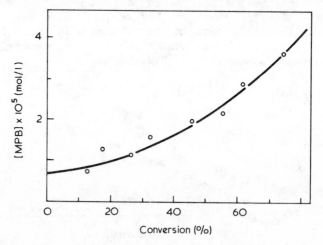

Fig. 11. Variation of MPB concentration with % conversion. [4-MP-1]=2.0 mole/l; [VCl₃]= 18.5mmole/l; [Al(i-Bu)₃]=37.0mmole/l; temperature=30°C.

merization, corrected for decrease in monomer concentration, reaches a constant value by about 20% conversion, and thus it seems reasonable to suppose that the number of active centers has also reached a constant value at this point. The continued increase, however, in the MPB concentration beyond this point, clearly indicates the formation of a non-propagative metal-bonded species. This is indicative of chain transfer with metal alkyl as has been proposed by Natta and Pasquon[17]. Similar observations have been made by other workers[18-21]. The initial non-linear increase in the MPB concentration, when plotted *versus* time, corresponds to the initial formation of active centers. This is followed by a long period of linear increase in the MPB concentration due to the transfer reaction. The slope of the linear portion of the graph is equal to the rate of chain transfer with aluminium alkyl (R_{ta}) which is seen to remain constant.

Since the polymerization reaction is characterized by an initial settling period it was felt necessary to evaluate the active center concentration (C_o) from the dependence of MPB concentration on conversion rather than time. The non-linearity of the [MPB] *versus* conversion plot is due to the depletion in monomer concentration with corresponding decrease in rate. The evaluation, however, of C_o is simplified by use of the first order plot of [MPB] *versus* $\log_{10}(100-\%$ conversion) which gives rise to a linear plot over the investigated range of conversion. Some typical results for the catalyst system $VCl_3/Al(i-Bu)_3$ are shown in Fig. 12. C_o is evaluated from

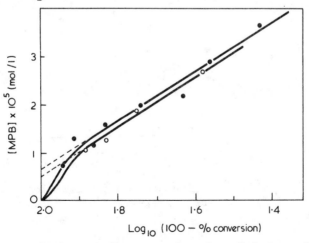

Fig. 12. Variation of MPB concentration with $\log_{10}(100-\%$ conversion) for $Al(i-Bu)_3$. [4-MP-1]=2.0 mole/l; [VCl_3]= 18.5mmole/l; [Al(i-Bu)_3]=37.0mmole/l. ●, 30°C; ○, 50°C.

the value of the intercept at zero conversion. Similar plots
are obtained using Al(i-Bu)$_3$ and Al(n-C$_{10}$H$_{21}$)$_3$, and these are
shown in Fig. 13 and 14. The corresponding plot when AlEt$_3$
is used is non-linear. This effect is most likely to arise
because of deactivation of the active centers — an effect
which was only observed in the case of AlEt$_3$. Values of C_0
together with corresponding values of R_{ta} for the different
aluminium alkyls which were used are summarized in Table 5.

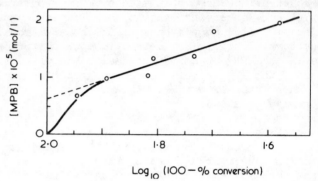

*Fig. 13. Variation of MPB concentration with log (100-% con-
version) for Al(n-Bu)$_3$. [4-MP-1]=2.0 mole/l; [VCl$_3$]=
18.5mmole/l; [Al(n-Bu)$_3$]=37.0mmole/l; temperature=30°C.*

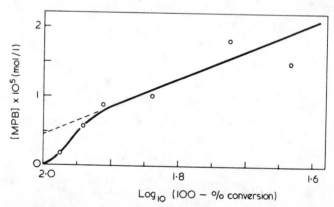

*Fig. 14. Variation of MPB concentration with log$_{10}$(100-% con-
version) for Al(n-Hex)$_3$. [4-MP-1]=2.0 mole/l; [VCl$_3$]=
18.5 mole/l; [Al(n-Hex)$_3$]=37.0mole/l; temperature=30°C.*

TABLE 5.

Variation in C_o and R_{ta} with aluminium alkyl.[a]

Aluminium alkyl	Temp. (°C)	$C_o \times 10^4$ (mole/mole VCl_3)	$R_{ta} \times 10^6$ (mole/l/min/VCl_3)
$AlEt_3$	30	6.10	17.2
$Al(i\text{-}Bu)_3$	30	3.78	3.24
$Al(i\text{-}Bu)_3$	50	2.90	12.8
$Al(n\text{-}Bu)_3$	30	3.30	1.53
$Al(n\text{-}C_6H_{13})_3$	30	2.30	0.87

[a][4-MP-1]=2.0mole/l; [VCl_3]=18.5mmole/l; [AlR_3]=37.0mmole/l; solvent = benzene.

It can be seen that the number of active centers in the polymerization system 4-MP-1/VCl_3/AlR_3 at 30°C lies in the range 2.3 - 6.1 x 10^{-4} mole/mole VCl_3 for the series of aluminium alkyls. These values compare with values of 3 - 10 x 10^{-3} mole/mole $TiCl_3$ found by several authors(17,22,23) for the system α-$TiCl_3$/$AlEt_3$/propylene.

This disparity in the values of active center concentration is capable of several explanations, the most important of which may be probably concerned with differences in the transition metal halide which has been used. It is quite conceivable that the number of active sites for VCl_3 would not be the same as for α-$TiCl_3$ because of differences in crystal structure and in stabilities of the different organo-transition species which would be involved. In addition several authors(10,24,25) observe that the rate of polymerization, and hence C_o, is dependent on the initial surface area of the transition metal halide. The initial surface area of the VCl_3 used in this study was 2.3 m^2/g, as measured by the Brunauer-Emmett-Teller (B.E.T.) method, which is low compared with a value of about $7m^2/g$ for the α-$TiCl_3$ used by other workers in active center determination. More recent experiments(26) show that when samples of ground VCl_3, having a surface area 34.5 m^2/g are used higher rates of polymerization for the catalyst system VCl_3/$Al(i\text{-}Bu)_3$ are obtained which correspond to values for C_o of 5.5 x 10^{-4} mole/mole VCl_3.

Kern *et al.*(27) have further shown that the number of active centers is dependent on the nature of the monomer, e.g., C_0 decreases from 4.5×10^{-3} to 3.8×10^{-3} mole/mole $TiCl_3$ on changing from propylene to butene-1.

These values for C_0 together with the known surface area of the VCl_3 sample raise some interesting queries concerning the location of the active centers. Cossee(28) has proposed that active site formation occurs at chlorine vacancies on the lateral faces of the crystal, a suggestion which is apparently supported by electron microscopy. More recently, however, Chirkov and Kissin(22) have suggested that active sites are also formed on the basal planes of the crystal lattice so as to cover the entire catalyst. The maximum number of active centers in the present system may be easily calculated. Thus assuming that the cross-sectional area of a poly(4-methylpentene-1) molecule is about 70 $Å^2$, i.e., the area likely to be taken up by a growing polymer chain, and also assuming that the surface area of the VCl_3 is 2.3 m^2/g, then the number of active sites corresponding to complete coverage of the crystal surface is about 9×10^{-4} mole/mole VCl_3, which is close to the maximum experimentally determined value of 6.1×10^{-4} mole/mole VCl_3. Since the lateral faces of the VCl_3 crystal may only comprise about 5% of the total surface area, a twenty-fold increase in surface area during polymerization would be necessary to accomodate all the active sites on these crystal faces. This calculation does not take into account any significant change in area through breakdown of the catalyst particles as has been suggested by Natta(17), nor does it take into account penetration of a solid catalyst structure, which is only loosely held together, by the various components present in the polymerization system. It neverthe-less, seems likely that active sites may be located on both the basal and lateral faces of the crystal lattice, but addi-tional experimentation is necessary before this situation can be further clarified.

5) *METAL-POLYMER BOND ANALYSIS*

5.1) Variation of Metal-Polymer Bond Concentration with Metal Alkyl Concentration.

The variation in MPB concentration with tri-isobutyl-aluminium concentration at constant monomer and constant vana-dium trichloride concentration for the catalyst system VCl_3/ $Al(i$-Bu$)_3$ is shown in Fig. 15. These experiments were carried out to a constant reaction time of 300 min. It is immediately apparent that the rate of formation of metal-polymer bonds,

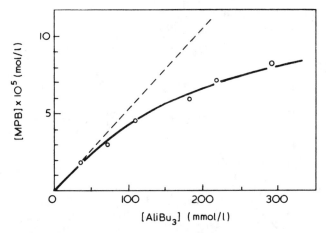

Fig. 15. Dependence of [MPB] on [Al(i-Bu)₃]. [4-MP-1]=2.0 mole/l; [VCl₃]=18.5mmole/l; temperature=30°C.

and thus the rate of chain transfer with metal alkyl (R_{ta}) is not directly proportional to the metal alkyl concentration in the solution. If chain transfer is with adsorbed metal alkyl then the rate of chain transfer is given by the equation:

$$R_{5a} = k_a \theta_A C_o \qquad [16]$$

where k_a is the rate constant for chain transfer with metal alkyl.

Rearrangement of eq. 16 and substitution for θ_A gives:

$$\frac{C_o}{R_{ta}} = \frac{1}{k_a} \left(\frac{1 + K_M[M] + K_A[A]}{K_A[A]} \right) \qquad [17]$$

Thus if the rate of transfer is with adsorbed metal alkyl then a plot of C_o/R_{ta} *versus* $[A]^{-1}$ should be linear. One problem in effecting such a plot is that in reality C_o is not constant throughout the duration of the polymerization reaction due to the 'settling' period. Consequently it is necessary to use the integral value of $C_o (\int_0^t C_o dt)$ over the reaction period. The value of $\int_0^t C_o dt$ over the reaction period is evaluated from the area under the steady state rate of polymerization (R_p) *versus* time plots of the type depicted in Fig. 16.

C_o under 'steady state' conditions, i.e., where R_p is constant has already been determined, and $\int_0^t C_o dt$ may now be simply derived from the relationship:

$$\int_0^t C_o dt = C_o (\text{Steady state}) \times \frac{\text{shaded area}}{\text{area ABCD}} \qquad [18]$$

The steady state value of C_o is found to be 7.0×10^{-6} mole/l under the experimental conditions employed, at an Al:V ratio of 2:1. This corresponds to a value of $\int_0^t C_0 \, dt$ equal to 4.85 x 10^{-6} mole/l for t = 300 min.

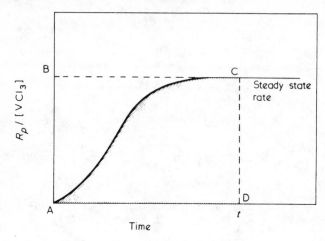

Fig. 16. Typical plot of rate of polymerization as a function of time.

One added complication, which is evident from experiments in which the Al:V ratio is increased by decreasing the vanadium trichloride concentration alone, is that the steady state rate is not absolutely constant, but only approaches a limiting value. Hence there is a slow increase in the steady state rate as the Al:V ratio is increased. Hence values of C_0 for different Al:V ratios $(C_{0(Al:V)})$ were calculated from the value of $C_{0(2:1)}$ and the steady state values of the polymerization rates by means of the relationship:

$$C_{0(Al:V)} = C_{0(2:1)} \times \frac{R_{p(Al:V)}/[VCl_3]}{R_{p(2:1)}/[VCl_3]}$$ [19]

Relevant steady state rate of polymerization *versus* time plots are shown in Fig. 17.

The rate of transfer using tri-isobutylaluminium at a catalyst ratio of 2:1($R_{ta(2:1)}$) has already been found to be 6.00 x 10^{-8} mole/l/min at an active center concentration of 7.00 x 10^{-6} mole/l for the catalyst system VCl$_3$/Al(i-Bu)$_3$. Thus the value of $\int_0^t C_0 dt$, equal to 4.85 x 10^{-6} mole/l calculated for a catalyst ratio of 2:1, corresponds to an integral rate of transfer of 4.16 x 10^{-8} mole/l/min.

The rates of chain transfer for other catalyst ratios were then found from the relationship:

$$R_{ta(Al:V)} = R_{ta(2:1)} \times \frac{[MPB]_{(Al:V)} - C_{o(Al:V)}}{[MPB]_{(2:1)} - C_{o(2:1)}} \quad [20]$$

where $[MPB]_{(Al:V)}$ and $[MPB]_{(2:1)}$ are the metal-polymer bond concentrations at $t = 300$ min at the respective catalyst ratios.

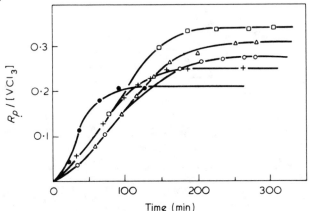

Time (min)

Fig. 17. Variation of rate of polymerization with time for different VCl_3 concentrations. [4-MP-1]=2.0 mole/l; [Al(i-Bu₃)]=37.0mmole/l. Temperature=30°C. [VCl₃] in mmole/l; □, 3.51; Δ, 5.44; O, 8.61; +, 18.2; ●, 44.3

A plot of $\int C_o/R_{ta}$ versus $1/[Al(i-Bu)_3]$ as shown in Fig. 18 is linear, with a positive intercept, thus confirming that the adsorption of metal alkyl may be described by a Langmuir type isotherm. Under these conditions:

$$\text{intercept} = (k_a)^{-1} \quad [21]$$

$$\frac{\text{slope}}{\text{intercept}} = \frac{1 + K_M[M]}{K_A} \quad [22]$$

Thus k_a may be evaluated directly whilst K_A and K_M require further data.

Similar linear plots are shown for triethylaluminium, tri-*n*-butylaluminium and tri-*n*-hexylaluminium and are depicted in Figs. 19 and 20. Values of the corresponding chain trans fer constants are recorded in Table 6.

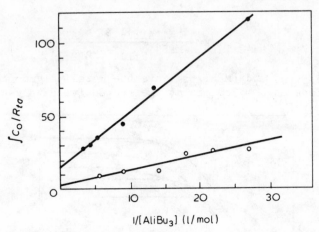

Fig. 18. *Langmuir plot for Al(i-Bu)₃. [4-MP-1]=2.0 mole/l; [VCl₃]=18.5mmole/l. ●, 300 min; ○, 70 min.*

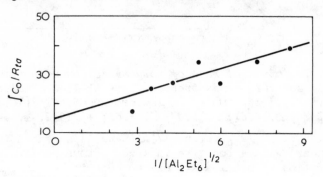

Fig. 19. *Langmuir plot for AlEt₃. [4-MP-1]=2.0 mole/l; [VCl₃]=18.0mmole/l; temperature=30°C; time=200 min.*

TABLE 6.

Chain transfer constants at 30°C.[a]

Aluminium alkyl	k_a (min^{-1})
AlEt$_3$	0.067 ± 0.003
Al(i-Bu)$_3$	0.067 ± 0.003
Al(n-Bu)$_3$	0.050 ± 0.005
Al(n-C$_6$H$_{13}$)$_3$	0.033 ± 0.001

[a][4-MP-1]=2.0mole/l; [VCl₃]=18.5mmole/l; solvent = benzene.

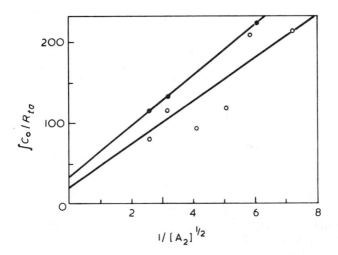

Fig. 20. Langmuir plots for Al(n-Bu)₃ and Al(n-Hex)₃. [4-MP-1]=2.0 mole/l; [VCl₃]=18.5mmole/l; ○, Al(n-Bu)₃, 270 min; ●, Al(n-Hex)₃, 360 min; temperature=30°C.

5.2) <u>Variation of Metal-Polymer Bond Concentration with Monomer.</u>

It would be expected that for experiments carried out at constant catalyst concentration, the concentration of metal-polymer bonds would be independent of the monomer concentration since:

$$[MPB]_t = C_0 + \int_0^t k_a \theta_A C_0 dt \qquad [23]$$

and since the fraction of surface covered by adsorbed metal alkyl (θ_A) varies only slightly over the investigated range. However, a plot of $[MPB]$ *versus* [4-MP-1] shows this not to be the case. This is evident from an examination of Fig. 21, which relates to experiments carried out to a constant reaction time of 300 min. This behaviour presumably arises from the slow initiation process(29).

The results of these experiments may be interpreted to provide data for the evaluation of K_M and K_A. A graph of specific activity of polymer *versus* $[M]^{-1}$ should be linear(11) with a positive intercept. Such a plot for tri-isobutylaluminium is shown in Fig. 22 and for this plot:

$$\frac{\text{slope}}{\text{intercept}} = \frac{1 + (1 + k_a t)K_A[A]}{K_M} \qquad [24]$$

177

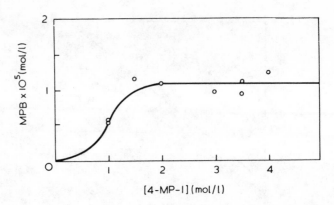

Fig. 21. Variation of MPB concentration with monomer concentration. [VCl₃]=18.5mmole/l; [Al(i-Bu)₃]=37.0mmole/l; temperature=30°C.

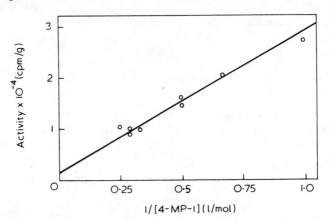

Fig. 22. Plot of activity versus 1/[4-MP-1]. [VCl₃]=18.5 mmole/l; [Al(i-Bu)₃]=37.0mmole/l; temperature=30°C.

Eq. 24, can be used in conjunction with eq. 22 to evaluate both K_A and K_M. On solving these equations the following values for K_M and K_A are obtained for the catalyst system $VCl_3/Al(i\text{-}Bu)_3/4\text{-}MP\text{-}1$

$$K_M = 0.164 \pm 0.020 \text{ 1/mole at } 30°C.$$

$$K_A = 5.12 \pm 0.20 \text{ 1/mole at } 30°C$$

Using the value $K_M = 0.164$ and K_A or $K^{\frac{1}{2}}K_A$ values may be deduced for other aluminium alkyls. These values are listed in Table 7.

TABLE 7.

Values of K_A for some aluminium alkyls at 30°C.

Aluminium alkyl	$K^{1/2}K_A$	K_A
Al$(i-Bu)_3$	----	5.12 ± 0.02
AlEt$_3$	7.83 ± 0.40	224 ± 10
Al$(n-Bu)_3$	0.97 ± 0.10	----
Al$(n-C_6H_{13})_3$	1.21 ± 0.06	----

The value of K_M = 0.164 ± 0.020 1/mole for the adsorption of 4-MP-1 onto VCl$_3$ at 30°C compares favourably with a value of 0.163 1/mole obtained by Vesely[30] for the adsorption of propylene onto TiCl$_3$ at 50°C.

5.3) Variation of Metal-Polymer Bond Concentration with Vanadium Trichloride Concentration.

This series of experiments, which was carried out to a constant reaction time of 240 min, and at constant monomer concentration, should be described[11] by an equation of the form:

$$[MPB]_t = P[A] + \frac{P k_a t K_A[A]^2}{1 + K_A[A] + K_M[M]} \qquad [25]$$

If an average value of 1.48 is assumed for the expression $(1 + K_A[A] + K_M[M])$ for these experiments, as this expression remains approximately constant over the concentration range employed, eq. 25 becomes:

$$[MPB]_t = P([A] + 56[A]^2) \qquad [26]$$

The plot of [MPB] *versus* $([A] + 56[A]^2)$ is linear (Fig. 23) with slope P equal to $(1.37 ± 0.03) \times 10^{-4}$, where P was defined by the experssion:

$$C_o = P[A] \qquad [27]$$

and for this series $[A] = 2[VCl_3]$, thus:

$$C_o = (2.74 ± 0.06) \times 10^{-4} \text{ mole/mole VCl}_3$$

This value of C_o compares with the value of $(3.78 ± 0.15) \times 10^{-4}$ which has already been obtained. The apparent discrep-

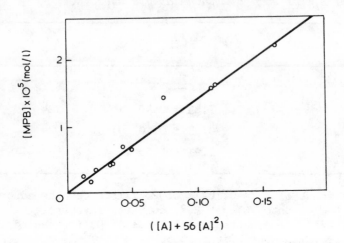

Fig. 23. Plot of [MPB] versus ([A] + 56[A]²). [4-MP-1]=2.0 mole/l; [Al(i-Bu)₃]: VCl₃ =2:1; temperature=30°C; time= 240 min.

ancy in these values is easily explained since the former value is the integral value of C_0 up to t equal to 240 min, whilst the latter is measured under steady state conditions. The integral value of C_0, calculated as described above, using the steady state value of C_0 of 3.78×10^{-4} mole/mole VCl₃, is found to be 2.68×10^{-4} mole/mole VCl₃, and compares very closely with the value of 2.74×10^{-4} mole/mole VCl₃ as determined above.

6) KINETIC ANALYSIS

6.1) Determination of the Rate Constant for Chain Propagation.

When propagation is considered to occur with adsorbed monomer the overall rate of polymerization is given by eq. 10. The value of Θ_M is, however, dependent on the nature of the metal alkyl, and consequently it is necessary to plot R_p against $\Theta_M C_0$ rather than against C_0 alone. An appropriate plot is shown in Fig. 24 and is linear with slope $k_p = (3.0 \pm 0.5) \times 10^3$ min⁻¹. It is thus seen that k_p is indepen-

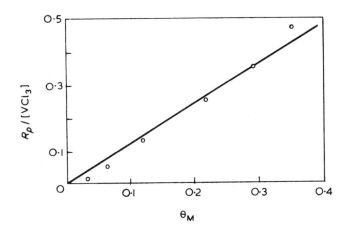

Fig. 24. Variation in rate of polymerization with Θ_M. [4-MP-1]=0.25 to 4.00mole/l; [VCl$_3$]=18.5mmole/l; [Al(i-Bu)$_3$] :[VCl$_3$]=2:1; temperature=30°C.

dent of the nature of the aluminium alkyl, a fact which strongly suggests that the species active in polymerization in the present study is an alkylated vanadium entity. These conclusions are not completely unambiguous since the apparent propagation rate constant might also be expected to be independent of the metal alkyl if the rate–determining step were the prior co-ordination of the monomer at the transition metal site(31,32). This is considered to be unlikely(33).

6.2) Variation of Rate of Polymerization with Alkyl Concentration.

It was suggested earlier that the decrease in the steady state rate observed with increase in metal alkyl concentration above an Al:V ratio of 2:1 was due to competitive adsorption of metal alkyl with monomer. Under these conditions the rate of polymerization is given by eq. 10. C_0 may be determined for different Al:V ratios as described earlier. A knowledge of K_A and K_M now allow Θ_M to be calculated. Calculated and experimental results are reported in Table 8. The

TABLE 8.

Comparison of calculated and theoretical rates at 30°C.[a]

$[Al(i-Bi)_3] \times 10^3$ (mole/l)	Experimental $R_p/[VCl_3]$ (mole/l/min/ mole VCl_3)	Θ_M	$\dfrac{C_o(Al:V)}{C_o(2:1)}$	Calculated $R_p/[VCl_3]$ (mole/l/min/ mole VCl_3)
33.7	0.274	0.216	1.00	0.260
66.6	0.266	0.196	1.14	0.266
110	0.234	0.173	1.20	0.250
138	0.233	0.161	1.23	0.240
217	0.186	0.131	1.30	0.211
289	0.176	0.117	1.34	0.187
369	0.164	0.102	1.38	0.167
491	0.140	0.086	1.39	0.142
558	0.142	0.079	1.40	0.130

[a] $[4-MP-1]=2.0$mole/l; $[VCl_3]=18.5$mmole/l; solvent = benzene.

agreement between the predicted and experimental values is good, deviations being within the limits of experimental error.

6.3) Variation of Rate of Polymerization with Monomer Concentration.

The steady state rate of polymerization is described by eq. 10, and consequently under conditions where C_o remains constant, the rate of polymerization should be directly proportional to Θ_M, i.e., a plot of R_p against Θ_M should be linear and pass through the origin. Θ_M at varying monomer concentrations can be evaluated from eq. 8 since both K_M and K_A are known. Fig. 25 shows a plot of Rp *versus* Θ_M as the monomer concentration is varied. This plot, which shows some

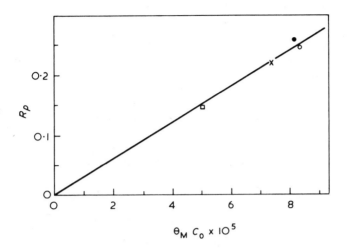

Fig. 25. Dependence of k_p on the nature of the metal alkyl.
[4-MP-1]=2.0 mole/l; [VCl$_3$]=18.5mmole/l; [AlR$_3$]=37.0
mmole/l; temperature=30°C. ●,Al(i-Bu)$_3$; ○, AlEt$_3$; x,
Al(n-Bu)$_3$; □, Al(n-Hex)$_3$.

scatter, deviates from linearity at low values of Θ_M (at low
monomer concentration). It has, however, been observed that
the number of active centers is reduced at low monomer concentrations and consequently C_0 is not constant in this
region.

A value of k_p equal to 3.10×10^3 min^{-1} is obtained from
the slope of the plot (for C_0 equal to 3.78×10^{-4} mole/mole
VCl$_3$). This value is in good agreement with the value
determined previously.

7) MOLECULAR WEIGHT ANALYSIS

The following termination and transfer reactions may be
formulated(5,34):

(a) Spontaneous terminations

$$\text{Cat-CH}_2\text{-CH}\text{\leavevmode\unskip}\hspace{-0.2em}\wedge\hspace{-0.4em}\wedge\hspace{-0.4em}\wedge\text{P} \xrightarrow{k_s} \text{Cat-H} + \text{CH}_2\text{=C}\wedge\hspace{-0.4em}\wedge\hspace{-0.4em}\wedge\text{P}$$

with R substituents below each.

This reaction may be followed by realkylation of the resulting
catalyst-hydride bond so that there is no change in the over-

all number of active centers, i.e.,

$$\text{Cat-H} + \text{CH}_2=\text{CHR} \xrightarrow{k_i} \text{Cat--CH}_2\text{-CH}_2\text{R}$$

There is no evidence that such a reaction is important at temperatures between 30°C and 40°C.

(b) Transfer to monomer:

$$\text{Cat-CH}_2\text{-}\underset{\underset{R}{|}}{\text{CH}}\text{\small\sim\sim\sim}P + \text{CH}_2=\text{CHR} \xrightarrow{k_m} \text{Cat-CH}_2\text{-CH}_2\text{R} + \text{CH}_2=\underset{\underset{R}{|}}{\text{C}}\text{\small\sim\sim\sim}P$$

(c) Transfer to tri-isobutylaluminium:

$$\text{Cat-CH}_2\text{-}\underset{\underset{R}{|}}{\text{CH}}\text{\small\sim\sim\sim}P + \text{Al}(i\text{-Bu})_3 \xrightarrow{k_a} \text{Cat-CH}_2\text{-}\underset{\underset{\text{CH}_3}{|}}{\text{CH}}\text{-CH}_3 +$$

$$\text{Al}(i\text{-Bu})_2\text{CH}_2\text{-}\underset{\underset{\text{CH}_3}{|}}{\text{CH}}\text{\small\sim\sim\sim}P$$

It will be realized that these equations are formulations only, in which k_s, k_m and k_a are rate constants for the spontaneous termination of growing chains, transfer with monomer and transfer with tri-isobutylaluminium respectively.

When chain transfer is with adsorbed metal alkyl, the rate of chain transfer is given by eq. 16. Similarly, for chain transfer with adsorbed monomer the rate of chain transfer is given by:

$$R_{tm} = k_m \theta_M C_o \qquad [28]$$

where k_m is the rate constant for chain transfer with adsorbed monomer.

If the propagation reaction is considered to occur between an adsorbed monomer molecule and an active center, and if the chain transfer reactions are with adsorbed monomer and adsorbed aluminium alkyl, the following equations are valid for the present system:

$$\overline{P}_n = \frac{\int_o^t k_p \theta_M C_o dt}{C_o + \int_o^t k_m \theta_M C_o dt + \int_o^t k_a \theta_A C_o dt} \qquad [29]$$

Integration and inversion yields:

$$\frac{1}{\overline{P}_n} = \frac{k_a \theta_A}{k_p \theta_M} + \frac{1}{k_p \theta_M t} + \frac{k_m}{k_p} \qquad [30]$$

Substitution for θ_M and θ_A gives:

$$\frac{1}{\overline{P}_n} = \frac{k_a K_A [A]}{k_p K_M [M]} + \frac{1/t}{k_p K_M [M]} + \frac{1/t(K_A [A])}{k_p K_M [M]} + \frac{(1/t) + k_m}{k_p} \qquad [31]$$

A plot of $(\overline{P}_n)^{-1}$ *versus* $[M]^{-1}$ for the present catalyst system is indeed linear as is shown in Fig. 26 and 27.

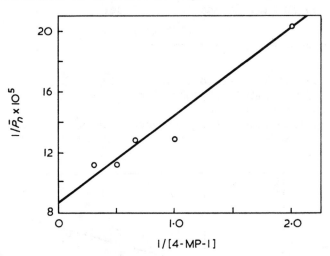

Fig. 26. Plot of $1/\overline{P}_n$ against $1/[4-MP-1]$. Al:V ratio=2.0:1; $[VCl_3]$=18.0mmole/l; temperature=30°C; solvent=benzene.

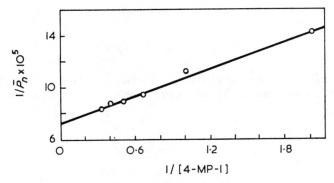

Fig. 27. Plot of $1/\overline{P}_n$ against $1/[4-MP-1]$. Al:V ratio=2.0:1; $[VCl_3]$=8.8mmole/l; solvent=benzene; temperature=40°C.

It is now possible to evaluate the rate constant for chain transfer with adsorbed monomer (k_m) since in accordance with eq. 31 a plot of $(\overline{P}_n)^{-1}$ *versus* $[M]^{-1}$ in linear with:

$$\text{intercept} = \frac{(1/t) + k_m}{k_p} \qquad [32]$$

In addition a plot of $(\overline{P}_n)^{-1}$ *versus* $[A]$ is also linear, as is shown in Fig. 28 and in this case:

$$\text{intercept} = \frac{1/t}{k_p K_M [M]} = \frac{(1/t) + k_m}{k_p} \qquad [33]$$

Using values of $k_p = (3.18 \pm 0.05) \times 10^3/\text{min}^{-1}$ at 30°C and $(14.6 \pm 2.5) \times 10^3/\text{min}^{-1}$ at 40°C, the values of k_m given in Table 9 can be calculated.

It is also possible(34) to check the value of k_m at 30°C from tritium end-group analysis and this value is also included in Table 9.

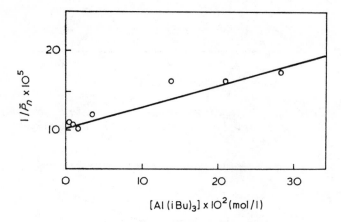

Fig. 28. Plot of $1/\overline{P}_n$ against $[Al(i\text{-}Bu)_3]$. $[4\text{-}MP\text{-}1]=2.0$ mole/ l; $[VCl_3]=8.7$ mmole/l; temperature=40°C; solvent=benzene.

TABLE 9.

Values of transfer constant with adsorbed monomer for the system $VCl_3/Al(i-Bu)_3/4-MP-1$.

Source	$k_m (min^{-1})$	
	30°C	40°C
Molecular weight data (equation [32]).	0.26 ± 0.06	
Tritium end-group data(34).	0.26 ± 0.08	
Molecular weight data (equation [32]).	0.26 ± 0.08	
Molecular weight data (equation [32]).		0.49 ± 0.10
Molecular weight data (equation [33]).		0.70 ± 0.15

8) *EFFECT OF ELECTRON DONORS*

The effect of a strong donor, triethylamine, which is capable of complex formation with both the transition metal and aluminium alkyl was investigated(35). Experiments were designed so as to vary the extent of the interaction of the donor with the catalyst components in an attempt to distinguish the relative importance of both complex formation with the transition metal and complex formation with the aluminium alkyl.

8.1) Effect of Order of Addition of Et_3N on the Course and Rate of Polymerization.

The effect of varying the order of addition on the steady state rate of polymerization under standard conditions is shown in Table 10. As can be seen the triethylamine was added in three distinct orders which proved to be of significance, viz:

187

$$VCl_3/Et_3N/4-MP-1/benzene/Al(\textit{i}-Bu)_3 \qquad \text{Order A}$$
$$VCl_3/4-MP-1/Et_3N/benzene/Al(\textit{i}-Bu)_3 \qquad \text{Order B}$$
$$VCl_3/4-MP-1/Al(\textit{i}-Bu)_3/benzene/Et_3N \qquad \text{Order C}$$

It is immediately apparent that under these conditions Et_3N has an activating effect, and that the order of addition is important.

TABLE 10.

Effect of order of addition of components on rate of polymerization.[a]

	Order of addition	$R_p/[VCl_3]$ (mole/l min. $[VCl_3]$)
Control[b]	$VCl_3/Al(\textit{i}-Bu)_3/Benzene/4-MP-1$	0.271
Control	$VCl_3/4-MP-1/Benzene/Al(\textit{i}-Bu)_3$	0.274
A	$VCl_3/Et_3N/4-MP-1/Benzene/Al(\textit{i}-Bu)_3$	0.292
B	$VCl_3/4-MP-1/Et_3N/Benzene/Al(\textit{i}-Bu)_3$	0.468
C	$VCl_3/4-MP-1/Al(\textit{i}-Bu)_3/Benzene/Et_3N$	0.467

[a] $[4-MP-1]=2.0$mole/l; $[VCl_3]=18.5$mmole/l; $[Al(\textit{i}-Bu)_3]=37.0$ mmole/l; $[Et_3N]=18.5$mmole/l (except for control runs); temperature$=30°C$. [b]Catalyst components aged for 30 min at $30°C$ and the monomer distilled in.

The course of polymerization in the presence of amine (orders A and B) is shown in Fig. 29, and is similar to that of the control polymerization, i.e., these is an initial settling period which is followed by a region during which the polymerization rate, corrected for decrease in monomer concentration, remains constant up to at least 55% conversion.

The key order of addition for catalyst activation is between monomer and amine. Thus the rates of polymerization for orders B and C, where monomer was added prior to amine, are almost double that of order A, where amine was added first. The polymerization activity is almost identical for

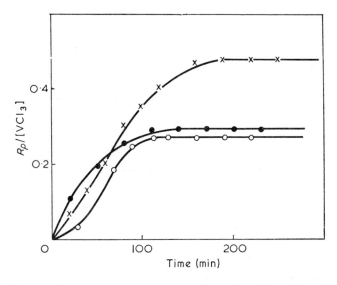

Fig. 29. Dependence of rate of polymerization on addition of
Et$_3$N. [4-MP-1]=2.0 mole/l; [VCl$_3$]=18.5mmole/l; [Al(i-
Bu)$_3$]=37.0mmole/l; temperature=30°C. O, Control poly-
merization: [Et$_3$N]=18.5mmole/l; ●, order of addition A;
x, order of addition B.

orders B and C where the order of addition of amine and metal
alkyl was varied, the monomer having been added beforehand.
These observations seem to suggest that there must be some
specific interaction between the monomer and VCl$_3$ surface
which is a prerequisite to active site formation.

8.2) Effect of Variation of Triethylamine Concentration on Rate of Polymerization.

For order A, when Et$_3$N is added prior to the monomer, the
polymerization rate is reduced when [Et$_3$N]: Al(i-Bu)$_3$=0.25,
as is shown in Table 11. This reduction is followed by slight
activation as the ratio is increased to 0.5. At equimolar
proportions of Et$_3$N and Al(i-Bu)$_3$ the steady state rate is,
however, reduced by a factor of 1000. This decrease in rate
is most likely associated with complex formation of the
aluminium alkyl which is complete in the presence of excess
amine(36). Alkylation of the transition metal is most likely
dependent on the presence of uncomplexed metal alkyl.

TABLE 11.

Dependence of rate of polymerization on triethylamine concentration.[a]

Order of addition	[Et$_3$N] (mmole/l)	$\dfrac{[Et_3N]}{[Al(i\text{-}Bu)_3]}$	$R_p/[VCl_3]$ (mole/l/min/mole VCl$_3$)
A	9.3	0.25	0.240
A	18.5	0.50	0.292
A	37.0	1.00	0.00023
B	9.3	0.25	0.423
B	18.5	0.50	0.468
B	27.8	0.75	0.684
Control	———	————	0.271

[a][4-MP-1]=2.0mole/l; [VCl$_3$]=18.5mmole/l; [Al(i-Bu)$_3$]=37.0mmole/l; temperature=30°C.

Polymerization in the presence of equimolar proportions of Et$_3$N and Al(i-Bu)$_3$ is quite distinct from that described previously, as is shown in Fig. 30. The polymerization rate increases to a maximum within 2 h and subsequently decreases over the next 200 h by a factor of about 12, after which the reaction continues at a slow but constant rate for at least a further 600 h. A detailed analysis(35) of the rate curve supports the hypothesis that two distinct polymerization systems are present, viz., a heterogeneous polymerization of steady rate and a homogeneous polymerization of decaying rate.

For order B, in contrast to order A, the steady state rate is increased significantly, even at low amine concentrations, when compared with that of the control experiment. This difference in behaviour between orders A and B again emphasizes the importance of the monomer–transition metal interaction. The lower rates for the order of addition A is believed to arise from a reduction in the number of active centers when amine is added prior to the monomer. The same mechanism of activation is considered to operate in both systems, but that this is not apparent for order A, as the rate is simultaneously lowered by reduction of active site concentration.

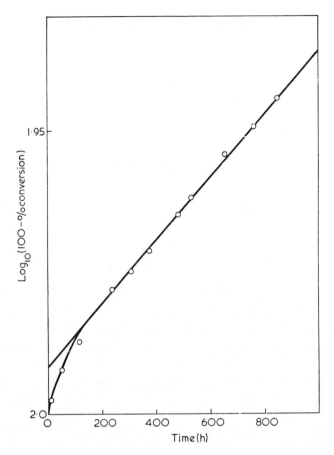

Fig. 30. Plot of log_{10}(100-% conversion) versus time. [4-MP-1]=2.0 mole/l; [VCl_3]=18.5mmole/l; [Al(i-Bu)_3]=37.0 mmole/l; [NEt_3]=37.0mmole/l; order of addition A; temperature=30°C.

8.3) Effect of Et₃N Addition During Polymerization on the Rate of Polymerization.

The effect of addition of varying amounts of Et₃N during the course of polymerization after the attainment of a steady rate, and therefore after attainment of a constant number of active centers is shown in Fig. 31. The effect of the addition of varying amounts of Et₃N is shown in Fig. 32. The rate activation is similar in magnitude to that observed when amine is added at the onset of polymerization (order B), and it is likely that the basic action of the amine is the same

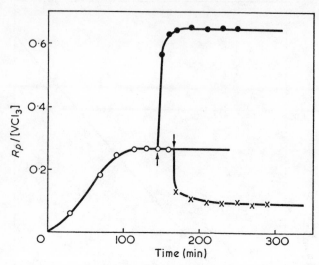

Fig. 31. *Effect of amine additions during course of polymer-*
ization. [4-MP-1]=2.0 mole/l; [VCl₃]=18.5mmole/l; [Al-
(i-Bu)₃]=37.0mmole/l; temperature=30°C. O, *Control*
polymerization. Polymerization after addition of tri-
ethylamine: O, [NEt₃]=18.5mmole/l; x, [NEt₃]=34.3mmole/
l. ↓ *denotes addition of donor.*

in both cases. The new activation, however, is achieved
rapidly within about 3-4 min of the addition, whereas (for
order B) initial site formation and chain initiation is complete only after about 100 min. Since this amine activation
is a much faster process than the initial formation of sites
active in polymerization, it is unlikely that these processe
are identical. Consequently, these findings are not compatible with proposals that activation is due to an increase
in the number of active centers caused by further breakdown
of the crystal lattice(37,38), or by increased site forming
capacity of the metal alkyl in the presence of amine(39,40).

These results can be rationalized in terms of the earlier
equations but in this case θ_M is now given by:

$$\theta_M = \frac{K_M[M]}{1 + K_M[M] + K_A[A] + K_D[D]} \qquad [34]$$

in the presence of additive D, assuming that only monomeric
species are adsorbed.

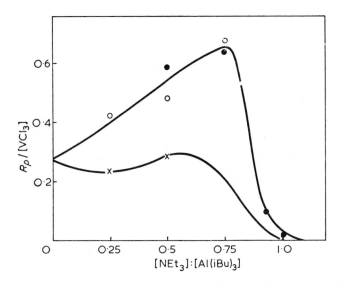

Fig. 32. Plot of $R_p/[VCl_3]$ versus $[NEt_3]:[Al(i-Bu)_3]$ ratio for different orders of addition of donor. [4-MP-1]= 2.0 mole/l; $[VCl_3]$=18.5mmole/l; $[Al(i-Bu)_3]$=37.0mmole/l; temperature=30°C. ✗*, Order of addition A;* O*, order of addition B;* ●*, donor added during polymerization.*

In order to quantitatively account for both the activating and deactivating effects of Et_3N in the present system the following interactions and equilibrium reactions must be considered:

$$Al(i-Bu) :NEt_3 \xrightleftharpoons{K_1} Al(i-Bu)_3 + NEt_3 \qquad [35]$$

$$VCl_3 + NEt_3 \xrightleftharpoons{K_D} VCl_3 \cdot NEt_3 \qquad [36]$$

$$VCl_3 + Al(i-Bu)_3 \xrightleftharpoons{K_A} VCl_3 \cdot Al(i-Bu)_3 \qquad [37]$$

$$VCl_3 + M \xrightleftharpoons{K_M} VCl_3 \cdot M \qquad [38]$$

When the metal alkyl concentration is in excess of the Et_3N concentration, then there will be little free Et_3N. In addition, the concentration of free metal alkyl will be reduced, as will its adsorption onto the VCl_3. Hence the value of Θ_M will increase and activation will be observed.

On the other hand, when amine is in excess, although the $K_A[A]$ term will be largely removed from the denominator of Θ_M, free amine is now present and will be strongly adsorbed and eq. 34 becomes:

$$\theta_M = \frac{K_M[M]}{1 + K_M[M] + K_D[D]_F} \qquad [39]$$

Under these conditions it can be shown(35) that:

$$R_p^{-1} = \frac{1 + K_M[M] + K[D]_F}{k_p C_o K_M[M]} \qquad [40]$$

A plot of R_p^{-1} *versus* [D] is shown in Fig. 33 and from the ratio of slope/intercept a value of $K_D = 360$ 1/mole is found. Triethylamine is thus very strongly adsorbed compared to $Al(i\text{-}Bu)_3$ ($K_A = 5.21/\text{mole}$).

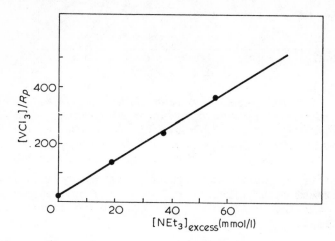

Fig. 33. *Plot of* $[VCl_3]/R_p$ *versus* $[NEt_3]$ *excess.* $[4\text{-}MP\text{-}1]=$ *2.0 mole/l;* $[VCl_3]=18.5 mmole/l;$ $[Al(i\text{-}Bu)_3]=37.0 mmole/l;$ *temperature=30°C.*

8.4) Number of Active Centers in Donor Modified Systems.

Active centers concentration was determined by the tritium quench technique, and results are shown in Table 12. It is apparent that, whereas the steady state rate for donor modified systems is over 150% higher, the number of active centers is only marginally higher and is in fact constant within the limits of experimental error. These results clearly confirm the proposal that donor activation occurs by modification of the adsorption equilibrium rather than an increase in the number of active centers.

TABLE 12.

Comparison of the rate of polymerization and number of active centers in the control and modified systems.[a]

	$R_p/[VCl_3]$ (mole/1/min/ mole VCl_3)	$C_0 \times 10^4$ (mole/mole VCl_3)
Control system	0.270	3.8 ± 0.4
Modified system[b]	0.684	4.6 ± 0.5

[a][4-MP-1]=2.0mole/1; [VCl₃]=18.5mmole/1; [Al(i-Bu)₃]=37.0 mmole/1; temperature=30°C. [b][NEt₃]=27.8 mmole/1 (order B).

In addition these results help to distinguish between two possible hypothetical models for the polymerization system: (i) The adsorbed species, monomer and alkyl, are mobile on the surface and polymer chains can grow at all possible centers. The momentary complexation of an alkyl molecule at any given center would hold up temporarily the propagation of a polymer molecule at the center - all centers would nevertheless carry polymer molecule; (ii) the alkyl when adsorbed would remain in position for periods of time, longer than the mean growth time of a polymer molecule. In this case polymer chains could only be present at centers where the alkyl was absent.

The results shown in Table 12, together with the kinetic results presented earlier which demonstrated that the addition of triethylamine during the course of the polymerization brought about a very rapid activation are considered to strongly favour the first hypothesis.

These studies demonstrate the validity of the proposed kinetic model in which propagation is considered to occur between an alkylated vanadium species and adsorbed monomer, and in which chain transfer is considered to occur with adsorbed monomer and adsorbed metal alkyl.

ACKNOWLEDGEMENT

The author wishes to acknowledge with gratitude the work of Drs. I.D. McKenzie and D.R. Burfield both in obtaining these results and in their interpretation. He also wishes to thank I.P.C. Business Press Ltd., for permission to use material and diagrams from publications in 'Polymer'.

LITERATURE CITED

(1) F. Eirich, and H.F. Mark, *J. Colloid Sci. 11*, 748 (1956).
(2) W.M. Saltman, *J. Polym. Sci. 46*, 375 (1960).
(3) H. Schnecko, M. Reinmoller, K. Weirauch, and W. Kern, *J. Polym. Sci.(C) 4*, 71 (1964).
(4) T. Keii, K. Soga, and N. Saiki, *J. Polym. Sci.(C) 16*, 1507 (1967).
(5) G. Natta, *J. Polym. Sci. 34*, 21 (1959).
(6) F.D. Otto, G. Parravano, *J. Polymer Sci.(A) 2*, 5131 (1964).
(7) G. Natta, F. Danusso, and I. Pasquon, *Colloq. Czech. Chem. Commun. 22*, 191 (1957).
(8) P.J.T. Tait, and I.D. McKenzie, IUPAC Symposium on Macromolecular Chemistry, Budapest, 1969, Preprints Vol. 2, p. 59.
(9) W.H. McCarty, and G. Parravano, *J. Polym. Sci.(A) 3*, 4029 (1965).
(10) K. Vesely, J. Ambroz, R. Vilim, and O. Hamrik, *J. Polym. Sci. 55*, 25 (1961).
(11) D.R. Burfield, I.D. McKenzie, and P.J.T. Tait, *Polymer 13*, 302 (1972).
(12) P. Cossee, *Tetrahedron Lett. 17*, 12 (1960).
(13) I.H. Anderson, G.M. Burnett, and P.J.T. Tait, *J. Polym. Sci. 56*, 391 (1962).
(14) D.R. Burfield, and P.J.T. Tait, *Polymer 13*, 315 (1972).
(15) A.S. Hoffman, B.A. Frier, and P.C. Condit, *J. Polym. Sci. C4*, 109 (1963).
(16) I.D. McKenzie, P.J.T. Tait, and D.R. Burfield, *Polymer 13*, 307 (1972).
(17) G. Natta, I. Pasquon, *Adv. Catalysis 11*, 1 (1959).
(18) J.C.W. Chien, *J. Amer. Chem. Soc. 81*, 86 (1958).
(19) C.F. Feldman, and E. Perry, *J. Polym. Sci. 46*, 217 (1960).
(20) A.D. Caunt, *J. Polym. Sci. C4*, 49 (1964).
(21) H. Schnecko, W. Lintz, and W. Kern, *J. Polym. Sci.(A-1) 5*, 205 (1967).
(22) Yu V. Kissin, S.M. Mezhikovsky, and N.M. Chirkov, *Eur. Polym. J. 6*, 267 (1970).
(23) R.W. Coover, J.E. Guillett, and F.S. Joyner, *J. Polym. Sci.(A-1) 4*, 2583 (1966).

(24) T. Keii, *Nature* *203*, 76 (1964).
(25) L. Kollar, A. Simon, and A. Kallo, *J. Polym. Sci.(A-1)* *6*, 937 (1968).
(26) J. Barberio, and P.J.T. Tait, (unpublished results).
(27) H. Schnecko, and W. Kern, I.U.P.A.C. Macromolecular Symposium, Budapest, 1969.
(28) P. Cossee and E.J. Arlman, *J. Catalysis* *3*, 99 (1954).
(29) D.R. Burfield, P.J.T. Tait, and I.D. McKenzie, *Polymer* *13*, 321 (1972).
(30) K. Vesely, *Pure Appl. Chem.* *4*, 407 (1962).
(31) A. Schindler, *J. Polym. Sci. B3*, 147 (1965).
(32) I. Pasquon, G. Natta, A. Zambelli, and A. Maringangelli, *J. Polym. Sci. C16*, 2501 (1967).
(33) P. Cossee, The Sterochemistry of Macromolecules (Ed. A.D. Ketley), Marcel Dekker, New York, 1967, p. 156.
(34) I.D. McKenzie, and P.J.T. Tait, *Polymer 13*, 510 (1972).
(35) D.R. Burfield, and P.J.T. Tait, *Polymer 15*, 87 (1974).
(36) G.A. Razuvaev, and A.I. Graevskii, *Dokl. Akad. Nauk, SSSR 128*, 309 (1959).
(37) J. Boor, Jr., *Macromol. Rev. 2*, 115 (1967).
(38) J. Ambroz, and O. Hamrik, *Coll Czech. Chem. Commun. 28*, 2550 (1963).
(39) R.L. McConnell, *J. Polym. Sci. A3*, 2135 (1965).
(40) H.W. Coover, Jr., and F.B. Joyner, *J. Polymer Sci. A3*, 2407 (1965).

Homogeneous Complex Catalysts of Olefin Polymerization

F.S. DYACHKOVSKII

Institute of Chemical Physics
USSR Academy of Sciences

After Ziegler's discovery of transition metal polymerization catalysts, a great number of analogous systems which
now represent quite a separate type of catalysts different
from those known earlier were suggested. In spite of the
fact that the mechanism of action of Ziegler catalysts has
been debatable up to the present, one can draw a number of
conclusions experimentally well-grounded and widely accepted
in the literature. The most definite results were obtained
in the course of studying homogeneous catalytic systems.
This paper is concerned mostly with the results that have
been obtained at the Institute of Chemical Physics in the
field of homogeneous catalytic systems. Kinetic studies and
application of various physical methods helped to define the
nature of some homogeneous catalysts active centres, to clear
up mechanism of their interaction with olefins, and to
receive quantitative characteristics of some elementary steps.
Among the homogeneous complex catalysts, two systems have
appeared to be studied in detail by the present. They are:
$(C_5H_5)_2TiCl_2$ + $AlEt_2Cl$ and $Ti(OR)_4$ + AlR_3 .

In 1957 $(C_5H_5)_2TiCl_2$ was announced to be a catalysts
component by the authors[1] who showed it to interact with
aluminium alkyls, forming a blue complex soluble in hydrocarbons. Though ethylene polymerization occurred in the presence of the complex, its activity appeared to be much less
compared to that of heterogeneous catalysts.

Almost at the same time Breslow and Newburg[2] showed
the complex to be considerably more active in the presence
of oxygen. That allowed them to conclude that active particles included titanium (IV) derivatives.

The $Ti(OC_4H_9)_4$ + $Al(C_2H_5)_3$ homogeneous catalytic system
was investigated by Bawn and Symocox[3]. The detailed study
of soluble complex catalysts based on titanium and vanadium
was made by Olive[4,5], Chirkov[6,7], Chien[8,9], Shilov[10,
11,12], Carrick[13], Bestian and Clauss[14].

For better understanding of mechanism of action of soluble polymerization catalysts, it is necessary first of all to identify the reaction products of their components.

1) PRODUCTS OF INTERACTION OF CATALYSTS COMPONENTS

1.1) Alkylation of Transition Metal.

It was shown by a spectrophotometic method(15) that interaction of $(C_5H_5)_2TiCl_2$ with AlR_2Cl resulted immediately in the formation of the $(C_5H_5)_2TiCl_2 \cdot AlR_2Cl$ complex followed by gradual alkylation of Ti. Kinetics of the alkylation reaction depends on the reaction conditions but it proceeds considerably faster compared to subsequent reactions (for example, titanium reduction). Activation energy of the alkylation reaction in toluene is approximately 14 kcal/mole. After the alkylation reaction the spectrum of the system in the visible region is completely identical to that of the $(C_5H_5)_2TiRCl \cdot AlRCl_2$ system.

The experimental evidence for the alkylation was obtained by studying the $(C_5H_5)_2TiCl_2 + AlMe_2Cl$ system with proton magnetic resonance(16). The authors showed that after mixing the components new resonance peaks could be observed in the PMR spectra, which corresponded to the protons of the following groups: $Al-CH_3$, $\tau=10.3$; $Ti-CH_3$, $\tau=9.0$; C_5H_5, $\tau=4.0$. The ratio of the proton band intensities of the $Ti-CH_3$ to C_5H_5 was 0.33 which is in agreement with the structure

$$(C_5H_5)_2Ti\!\!<^{\displaystyle CH_3}_{\displaystyle ClAlCH_3Cl_2}$$

It is noteworthy that the chemical shift of the $Ti-CH_3$ group in the presence of $Al(CH_3)_2Cl$ did not differ considerably from the proton chemical shift of the same group in $(C_5H_5)_2TiCH_3Cl$ synthesized separately. Hence we can conclude that the observed methyl group of titanium is not connected with an aluminium atom by a bridge bond; the formation of a bridging methyl group (for example, in trimethyl aluminium) would change the chemical shift of the methyl protons.

The evidence of titanium being alkylated in the $(C_5H_5)_2TiCl_2 + AlMe_2Cl$ system was obtained by mass spectrometry*. Stable bands corresponding to masses of the $(C_5H_5)_2TiCH_3$ and $(C_5H_5)_2TiCH_3Cl$ particles were discovered in the vapour mass

---

* The $(C_5H_5)_2TiCH_3Cl \cdot AlCH_3Cl_2$ complex mass spectra were recorded at the Kioto University, Japan by Dr. Ueno.

spectra of the $(C_5H_5)_2TiCH_3Cl \cdot AlCH_3Cl_2$ complex. Transition metal alkylation is a common reaction for all Ziegler-type complex catalysts. The reaction rate and degree of alkylation of a transition metal are determined by the nature and the character of a metalorganic compound. The alkylation of low valence transition metal derivatives can proceed via oxidative addition. For example, in the presence of an olefin:

$$\rangle Ti + CH_2=CH_2 + Ti\langle \longrightarrow \rangle Ti - CH_2 - CH_2 - Ti\langle \qquad [1]$$

or alkyl halides:

$$\longrightarrow Ti + RHal \longrightarrow \rangle Ti\langle{\substack{R \\ Hal}} \qquad [2]$$

The course of these reactions was studied in some works(17,18).

The participation of the transition metal – carbon bond in the catalytic reactions is now well-established and will be discussed further.

1.2) Electric Conductivity of Catalytic Complexes.

As early as 1960 Shilov supposed(10) the $(C_5H_5)_2TiRCl \cdot AlRCl_2$ complex to be in equilibrium with positive ions containing titanium:

$$(C_5H_5)_2TiRCl \cdot AlRCl_2 \rightleftarrows [(C_5H_5)_2TiR]^+ + [AlRCl_3]^- \quad [3]$$

$$(A)$$

It was found that after the addition of dicyclopentadienyl titanium dichloride, the electric conductivity of benzene solution of dimethyl aluminium chloride increased by several fold. A benzene solution of $(C_5H_5)_2TiCl_2$ without $AlMe_2Cl$ has no noticeable conductivity. This means that the $(C_5H_5)_2TiCH_3$ $-Cl \cdot AlCH_3Cl_2$ complex which was produced by interaction of $(C_5H_5)_2TiCl_2$ and $AlMe_2Cl$ dissociated into ions. The complex A has electric conductivity which is proportional to the square root of its concentration. The concentration of ions in the solution was not large and uncontrolled impurities effected the ion equilibrium to a great extent.

The direct evidence for the existence of alkylated titamium ions in this system was obtained by an electrodialysis method(19). The $(C_5H_5)_2TiCH_3Cl \cdot AlCH_3Cl_2$ complex solution was subjected to electrodialysis in dichloroethane. The number of ions containing titanium and migrating through a membrane to the cathode chamber was approximately 40% of the total ion current. Therefore, most of positive ions contain titanium atoms in the $(C_5H_5)_2TiCH_3Cl \cdot AlCH_3Cl_2$ complex solution. There were no titanium atoms among the negative ion. Electrodialysis of the $(C_5H_5)_2TiC^*H_3Cl \cdot AlC^*H_3Cl_2$ complex(20) showed that the labeled C^*H_3 appear together with titanium in the cathodic chamber, the C^*H_3/Ti ratio being close to unity. This evidence fully confirmed the dissociation of complex A according to eq. [3] and the structure of positive ions $[(C_5H_5)_2TiR]^+$.

The formation of the polar $Cl_2TiCH_3Cl \cdot AlCH_3Cl_2$ complex (A') in catalytic system $TiCl_4$ + $AlMe_2Cl$ was observed by Bestian and Clauss(21). Later in the work(22) the electric conductivity and electrodialysis of the complex formed were studied. The electric conductivity of the A' complex solutions increased approximately linearly with the square root of the initial titanium concentration value. An increase of the electric conductivity in the formation of the complex could be explained by its dissociation into ions according to the following scheme:

$$Cl_2TiCH_3Cl \cdot AlCH_3Cl_2 \rightleftharpoons [Cl_2TiR]^+ + [ClAlCH_3Cl_2]^- \quad [3']$$

In the course of complex A' electrodialysis the quantity of titanium in the cathodic chamber increased in direct proportion to the current. The absorption spectrum in the visible region of the solution in the cathodic chamber after electrodialysis corresponded to the spectrum of the complex A' solutions. All these data confirmed the dissociation of complex A' into ions according to the scheme [3'].

The formation of ions in other catalytic homogeneous systems of polymerization(23,24), oligomerization(25), and hydration(25) of olefins was observed by many authors. The participation of ionic species in catalysis was quite convincingly demonstrated for several systems.

1.3) Electron Paramagnetic Resonance of Catalyst Components' Reaction Products.

The complexes formed in the homogeneous catalytic systems $(C_5H_5)_2TiCl_2$ + AlR_2Cl were investigated by EPR in detail by A.E. Shilov(27,28). EPR was also used to study many other systems: $(C_5H_5)_2TiCl_2$ + AlR_2H (28); $(C_5H_5)Ti(OR)_3$ + AlR_3, $Al(CH_3)_3$; $(C_5H_5)TiCl_3$ + AlR_3, AlR_2Cl (29); VCl_4 + AlR_3; $VOCl_3$ + AlR_3 (30); $(C_5H_5)_2VCl_2$ + AlR_3, AlR_2Cl, $AlCl_2R$ (31); $TiCl_4$ + AlR_3 (32). All the results obtained could be summarized in the following way. The products of transition metal compounds reduction in the form of complexes with aluminium derivatives are formed by interaction of alkyls, aluminium alkyls halides with the derivatives of titanium(IV), titanium(III), vanadium(V), and vanadium(IV) at the expense of donor-acceptor or electron deficient bonds. The superhyperfine structure of the spectra helped to elucidate the structures of the complexes, in some cases rather definitely.

Despite a great number of complexes studied by EPR, there was no apparent relation between EPR signal intensity and catalytic activity. Thus, it would be incorrect to regard the EPR signal as an indicator of catalytic activity of complex catalysts. However, it is not excluded that a catalytically active particle in some systems can be paramagnetic.

1.4) Gaseous Products and Question of Free Radical Formation.

From the very beginning most investigators have come to a conclusion that the process of olefin polymerization by Ziegler catalysts was not a radical one. However, the following simple scheme was sometimes proposed,

$$TiCl_4 + AlR_3 \longrightarrow Cl_3TiR + AlR_2Cl$$

$$Cl_3TiR \longrightarrow TiCl_3 + R \cdot$$

[4]

Actually, this scheme was not confirmed experimentally and contradicted a number of experimental results.

By reacting aluminium alkyls with transition metal halides, products of disproportionation of aluminium alkyl groups are formed: ethane, ethylene in case of $AlEt_3$; isobutene and isobutylene in case of $Al(i-Bu)_3$, and so on.

In the homogeneous systems including $(C_5H_5)_2TiCl_2$ the reaction proceeds with a change in titanium valence by unity and at Al/Ti > 1 ratio ethane is formed in amounts nearly equal to half the titanium used (10). Ethylene is always formed in far smaller amounts due to polymerization or oxidative addition (for example, in case of the $Ti(OR)_4 + AlR_3$ system)(17). For the $TiCl_4 + AlEt_3$ system with a large excess of $AlEt_3$, the C_2H_6/Ti ratio reaches 1.5 indicating further reduction of titanium.

Changing the solvent from aliphatic to aromatic and addition of free radicals acceptors (e.g., anthracene) have no effect on the yield of alkane. In the course of the reaction of $Ti(OR)_4$ with $AlEt_3$ in perdeuterotoluene, $C_2H_5D(33)$ is not found in ethane formed.

All these results clearly indicated the absence of free alkyl radicals in the reaction of titanium compounds with aluminium alkyls. Really, if ethane and ethylene had been the products of disproportionation of free alkyl radicals, then butane should also be produced. However, butane was practically absent in the reaction products in the systems based on ethyl derivatives of aluminium. The quantity of alkane formed relative to the transition metal compound indicated no abstraction of hydrogen atom from the solvent, which was confirmed also by experiments with deuterated toluene.

Whereas, under such conditions free radicals would react with the solvent: R + DS ⟶ RD + S. Therefore, one can conclude that alkanes and olefins formed by interaction of transition metal derivatives with metal alkyls and metal alkyl halides are the products of non-radical disproportionation of alkyl groups(34,35,36).

2) *MECHANISM OF TRANSITION METAL REDUCTION.*

As it has been already pointed out, the first stage of interaction of complex catalysts components is the formation of an alkyl derivative of transition element. The alkyl derivatives formed closely relate to the initiation of polymerization. In most cases the reduction of a transition element occurs as a result of subsequent transformations of alkyl derivatives without free radical formation.

Kinetic investigations of the reaction of $(C_5H_5)_2TiCl_2$ with $AlEt_2Cl$ suggest that the reduction rate determining step is the decomposition of a dialkyl titanium derivative(10):

$$\begin{matrix} C_5H_5 & & C_2H_5 \\ & \diagdown Ti \diagup & \\ C_5H_5 & & C_2H_5 \end{matrix} \longrightarrow (C_5H_5)_2Ti + C_2H_6 + C_2H_4 \qquad [5]$$

The reaction evidently proceeds as intramolecular disproportionation through a five-membered cyclic activated complex

$$\begin{matrix} & & CH_2 - CH_2 \\ \diagdown Ti \diagup & & | \\ & R - H \end{matrix}$$

There is virtually no intramolecular combination of alkyl groups.

The authors of the work(37) have also come to a conclusion, when studying the reaction of vanadium compounds with aluminium alkyls, that the reaction proceeds through dialkyl vanadium derivatives. The vanadium reduction follows disproportionation of alkyl groups in the coordination sphere of the transition metal.

It would be interesting to note that disproportionation of alkyl groups in the coordination sphere of a transition metal was observed in some cases even for methyl derivatives. Thus the decomposition of $(CH_3)_2TiCl_2$ and $(CH_3)_4Ti(38)$ proceeded to a considerable extent according to the following reaction:

$$\begin{matrix} & CH_3 \\ \diagdown Ti \diagup & \\ & CH_3 \end{matrix} \longrightarrow \diagup Ti = CH_2 + CH_4 \qquad [6]$$

Small activation energy of this reaction is accounted for by the conjugation of carbene formed with a transition metal.

In addition to the intramolecular mechanism, bimolecular interaction of alkyl derivatives is also possible. For example, the reaction

$$2RTiCl_3 \longrightarrow R_{(+H)} + R_{(-H)} + 2TiCl_3 \qquad [7]$$

has been shown in some works(39,40).

The monomolecular decomposition resulting in olefin elimination and hydride formation is also possible:

$$\Big\rangle M - C_2H_5 \longrightarrow \Big\rangle MH + C_2H_4 \ . \qquad [8]$$

This is a reverse reaction of olefin insertion for the M-H bond which has been demonstrated for hydration of unsaturated compounds. The hydride formed must quickly react with alkyl derivatives forming C_2H_6.

The processes of transition metal reduction in catalytic systems are of great importance to polymerization as they terminate the growth of a polymer chain.

Interesting elementary reactions were observed when studying mechanism of titanium reduction in homogeneous catalytic system $Ti(OR)_4 + AlEt_3$. The considerable quantities of gas components produced in the interaction suggests that titanium is extensively reduced(41,42). At the Al/Ti > 2 ratio ethylene was not observed in the gas products. Close investigation of the reaction products in liquid phase showed(17,43) that the reaction of ethylene with the catalytic system is not limited to insertion in the Ti-C bond. It was found that ethylene also oxidized the titanium derivatives formed in the reduction to give compounds with fragments like $TiCH_2$-CH_2Ti. The general scheme may be presented in the following way:

$$Ti(OC_4H_9)_4 + AlEt_3 \longrightarrow Ti(OR)_3C_2H_5 + AlEt_2(OC_4H_9)$$
$$C_2H_5Ti(OR)_3 + AlEt_3 \longrightarrow Et_2Ti(OR)_2 + AlEt(OR)_2$$
$$(C_2H_5)_2Ti(OR)_2 \longrightarrow C_2H_6 + (OR)_2TiC_2H_4 \qquad [9]$$
$$2(RO)_2TiC_2H_4 \longrightarrow (RO)_2TiCH_2CH_2Ti(OR)_2 + C_2H_4$$

The considerable amount of dideuteroethane evolved after the hydrolysis of liquid phase by D_2O confirmed the formation of the $TiCH_2CH_2Ti$ groups. The methylene groups can react with metallo-organic compounds leading to further metallation of a hydrocarbon bridge:

$$\Big\rangle TiCH_2CH_2Ti\Big\langle + MR \longrightarrow \Big\rangle Ti\overset{\overset{M}{|}}{-}CH-CH_2-Ti\Big\langle + HR \qquad [10]$$

The compounds including the $\Big\rangle TiCH_2CH_2Ti\Big\langle$ group are, evidently, active in polymerization as the fraction of dideuteroethanes found after the hydrolysis by D_2O decreased in the presence of a monomer. Thus, oxidizing the low-valent transition metal derivatives by olefins may lead to the formation of metal-carbon bonds that can act as active centres of polymerization.

We may believe this to be the way by which active centres are produced in alkyl-free complex catalytic systems(44-46).

3) *MECHANISM OF POLYMERIZATION INITIATION*

The study of polymerization initiation and nature of the active centre in the homogeneous $(C_5H_5)_2TiCl_2$ + $AlEt_2Cl$ system is difficult because of the concurrent reduction reaction. More convenient for this purpose is the complex formed by interaction of $(C_5H_5)_2TiCl_2$ with $Al(CH_3)_2Cl$. Upon mixing these two components, a red solution is produced, the spectrum and catalytic activity of which do not change for many hours. As it was described above, both spectroscopic investigations and NMR spectra(15,16) of the complex obtained indicated the following structure:

$$(C_5H_5)_2Ti \underset{ClAlCH_3Cl_2}{\overset{CH_3}{\diagdown}}$$

The reduction of $(C_5H_5)_2TiCl_2$ with dimethyl aluminium chloride takes place only in the presence of olefins, the colouring of the solution changing from red to blue. In case of ethylene one observes the further formation of polyethylene. In contrast to ethylene which is absorbed in amounts considerably more than the 1 mole C_2H_4/1 mole Ti ratio, there is no polymerization for other α-olefins (propylene, butene, pentene, decene, etc.). Instead, the interaction proceeds mostly in accordance with the scheme:

$$2(C_5H_5)_2TiCH_3Cl \cdot Al(CH_3)Cl_2 + CH_2 = CHR \longrightarrow$$
$$2(C_5H_5)_2TiCl_2Al(CH_3)Cl + CH_2 = \underset{CH_3}{\overset{|}{C}} - R + CH_4 \qquad [11]$$

These results can be explained by the mechanism of titanium reduction(47,48). Alkyls capable of disproportionation in the coordination sphere of metal are formed by insertion of α-olefin:

$$\rightarrow Ti-CH_3 + CH_2 = CHR \longrightarrow \rightarrow TiCH_2 - \underset{}{\overset{CH_3}{\overset{|}{C}H}} - R$$

$$[12]$$

$$\underset{\diagdown CH_3}{\overset{\diagup CH_2-CH \diagdown R}{Ti}} \longrightarrow CH_4 + CH_2 = C \underset{\diagdown CH_3}{\overset{\diagup R}{}} + \diagdown Ti$$

The insertion of olefins in the Ti-C bond was established by PMR and spectrophotometric measurements(48,49). Optical spectra in the visible region showed complex to react with ethylene,

$$(C_5H_5)_2Ti\begin{cases} CH_3 \\ Cl \cdot Al(CH_3)Cl_2 \end{cases} + C_2H_4 \longrightarrow$$

$$(C_5H_5)_2Ti\begin{cases} C_3H_7 \\ Cl \cdot Al(CH_3)Cl_2 \end{cases}$$

[13]

prior to titanium reduction. PMR study of the $(C_5H_5)_2TiCH_3$-$ClAl(CH_3)Cl_2$ complex solutions in the presence of phenyl acetylene showed that the resonance of the Ti-CH_3 group disappeared whereas the one corresponding to the protons of the Al-CH_3 group did not change.

The difference in the behavior of ethylene as compared to other α-olefins is not difficult to understand from the point of view of non-radical mechanism of titanium reduction. When ethylene inserts in the Ti-CH_3 bond, rather stable non-branched alkyl groups appear on a titanium atom, the possibility of inserting new C_2H_4 molecules being preserved. In case of other α-olefins the alkyl group produced contains a weakly-bonded tertiary hydrogen atom:

$$\rightarrow Ti-CH_3 + CH_2=CHR \longrightarrow \rightarrow Ti-CH_2-\overset{\overset{\displaystyle R}{|}}{\underset{\underset{\displaystyle H}{|}}{C}}-CH_3$$

[14]

In this case the disproportionation (leading to titanium reduction and catalyst deactivation) occurs readily. This explains why, in contrast to ethylene, other α-olefins can not be polymerized by this catalytic system.

Changing the initial concentration of $(C_5H_5)_2TiCl_2$ changes the concentration of catalytic complex $(C_5H_5)_2TiCH_3ClAl(CH_3)$-$Cl_2$. Therefore, the initial rate of reaction of this complex with ethylene appears to be proportional to the square root of the titanium concentration:

$$-\frac{d[Ti^{+4}]}{dt} = k_{eff} [A]^{1/2}[C_2H_4]$$

[15]

The same kinetic equation is observed for the reduction of $Cl_2TiCH_3ClAlCH_3Cl_2$ in the presence of heptene.

4) *NATURE OF POLYMERIZATION ACTIVE CENTRE*

The spectrally measured reaction rate of ethylene inser-
tion in the Ti–CH$_3$ bond of $(C_5H_5)_2TiCH_3Cl \cdot AlCH_3Cl_2$ complex and
the rate of ethylene polymerization by this catalytic complex
depends strongly upon the environment. The rates increase
considerably with increase of the ion concentration in the
solution(10,50,51). In benzene–heptane mixtures of various
ratios, where the dielectric constant does not change much
yet the electric conductivity of the $(C_5H_5)_2TiCH_3ClAlCH_3Cl_2$
complex increases sharply with the increase of benzene content
in the mixture, the rate of ethylene polymerization and the
rate of reaction of the complex with α–olefins are approximately proportional to the electric conductivity (Fig. 1 and
Table 1).

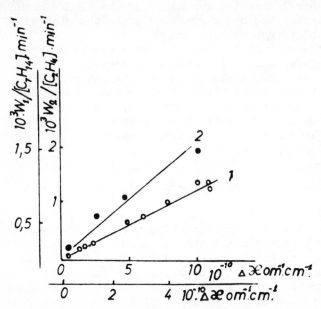

*Fig. 1. Variation of initial reaction rates
with changes in conductivity* Δ_K *at 20°: (1)
reduction in the presence of heptane, (2) in
ethylene polymerization. Lower line is rela-
tion of* w_1 *to* Δ_{K1}; $[(C_5H_5)TiCl_2]$ = $2.45 \cdot 10^{-3}$;
$[Al(CH_3)_2Cl]$ = $1.2 \cdot 10^{-2}$; $[C_7H_{14}]$ = $6.9 \cdot 10^{-2}$
mol/l; upper line is relation of w_2 *to* Δ_{K2};
$[(C_5H_5)_2TiCl_2]$ = $5 \cdot 10^{-3}$; $[Al(CH_3)_2Cl]$ = $2.5 \cdot$
10^{-2} *mol/l;* $P_{C_2H_4}$ = 438 mm Hg.

TABLE 1.

Electroconductivities and rates of decene reaction with complex A at 18°C.

Solvent and its composition (in mole %) of the first component	D	$\kappa \times 10^{11}$ ohm/cm.	$\tau^{1/2}$ min.
Heptane + benzene			
72	1.97	1.26	236
72	1.97	1.26	206
Benzene			
100	2.28	27	27
100		22	27
100		22	40
Dichloroethane + benzene			
7	2.5	107	30
13	2.8	320	20
37	3.6	1,720	15
Chlorobenzene + heptane			
93			14.0
93	5.4	13,200	12.5
Methylene chloride + benzene			
5	2.5	183	16.0
15	2.7	505	12.0
35	3.4	4,770	6.5
Methylene chloride + heptane			
41	3.1	1,570	8.0
41			8.0
Dichloroethane + heptane			
94			0.68
94	9.1	1,360,000	0.68

$[A] = 4.18 \ 10^{-3}$ mole/l; $[Al(CH_3)_2Cl] = 0.02$ mole/l; $[C_{10}H_{20}] = 4.2 \ 10^{-2}$ mole/l.

These results, together with the kinetic equation obtained, are in correspondence with the hypothesis(10) that free ions $[(C_5H_5)TiR]^+$ are the active species in ethylene polymerization and the reaction of complex $(C_5H_5)_2TiCH_3ClAlCH_3Cl_2$ with other olefins. In the ionic dissociation of the complex,

$$(C_5H_5)_2TiRCl \cdot AlRCl_2 \overset{K'}{\underset{}{\rightleftarrows}} [(C_5H_5)_2TiR]^+ + [AlRCl_3]^- \quad [16]$$

the concentration of the $[(C_5H_5)_2TiR]^+$ ions is given by

$$[(C_5H_5)_2TiR]^+ = (K')^{1/2} [Ti]^{1/2} \quad [17]$$

Thus, the kinetic eq. 15 can be explained if the rate is determined by a reaction of the active ion with an olefin molecule.

The kinetics of the $(C_5H_5)_2TiC_3H_7Cl \cdot AlCH_3Cl_2$(B) complex reactions(50) in various solvents was measured spectrophotometrically (Fig. 2). The decrease of the complex $(C_5H_5)_2Ti$-PrCl·AlCH$_3$Cl$_2$ concentration is related to the reduction of titanium to a blue complex of trivalent titanium. Figure 2 shows a strong dependence of the B complex formation rates on the solvent nature.

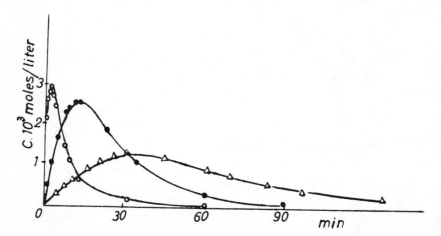

Fig. 2. Kinetic curves for complex B formation and decomposition in heptane (△), benzene (○), and chlorobenzene (●).

The general scheme of ethylene interaction with a catalytic system can be represented in the following way:

$$(C_5H_5)_2TiCH_3ClAl(CH_3)Cl_2 \rightleftarrows [(C_5H_5)_2TiCH_3]^+ +$$

$$[Al(CH_3)Cl_3]^-$$

for ethylene:

$$[(C_5H_5)_2TiCH_3]^+ + C_2H_4 \longrightarrow [(C_5H_5)_2TiC_3H_7]^+$$

$$[(C_5H_5)_2TiC_3H_7]^+ + nC_2H_4 \longrightarrow \text{polymerization}$$

$$(C_5H_5)_2Ti(CH_2CH_2R)Cl \cdot AlCH_3Cl_2 \rightleftharpoons (C_5H_5)_2Ti \Big\langle {CH_2CH_2R \atop CH_3} + AlCl_3$$

$$(C_5H_5)_2Ti \Big\langle {CH_2CH_2R \atop CH_3} \longrightarrow CH_4 + CH_2 = CHR + (C_5H_5)_2Ti \qquad [18]$$

$$(C_5H_5)_2Ti + [A] \longrightarrow \text{blue complex}$$

for α-olefin:

$$[(C_5H_5)_2TiCH_3]^+ + CH_2 = \dot{C}HR \xrightarrow{\text{slow}} (C_5H_5)_2Ti^+CH_2 - CH\Big\langle {R \atop CH_3}$$

$$(C_5H_5)_2Ti \Big\langle {R_1 \atop ClAlCH_3Cl_2} \xrightarrow{\text{rapid}} \text{blue complex}$$

This scheme is consistent with the data on electrodialysis results(20). The ethylene polymerization was shown to occur only in the cathodic (not anodic) chamber of a dialyzator,

However, the ionic mechanism is not general. It is possible that it is not the charge but the free orbitals and the absence of steric obstacles (e.g., for olefin polymerization) on an alkylated transition metal atom that is the necessary condition for catalytic activity. For instance, in the $Ti(OR)_4 + AlEt_3$ catalytic system both ions and unchanged complexes can be active centres. Its dimerization rate of ethylene to butene-1 (52,53) is not noticeably dependent on solvent polarity (and, therefore, electric conductivity of the system). In this case the dititanium compounds of the Ti-CH$_2$-CH$_2$-Ti type are active centres. The coordination of two ethylene molecules on titanium atoms followed by hydrogen atom transfer lead to butene formation:

$$\rangle TiCH_2CH_2Ti \langle \longrightarrow C_4H_8 + \rangle TiCH_2CH_2Ti \langle \qquad [19]$$
$$CH_2 \updownarrow CH_2CH_2 \updownarrow CH_2$$

The reaction rate is proportional to the dititanium compound concentration. In this case we deal with a purely coordination mechanism.

Also in the catalytic systems $(C_5H_5)_2TiEtCl + AlEt_2Cl$ and $(C_5H_5)_2TiEtCl + AlEtCl_2$ at $Al/Ti = 1$, free ions do not play an important role. However, the polar complex $(C_5H_5)_2$-$TiRCl \cdot AlRCl_2$ is much more active than the less polar complex $(C_5H_5)_2TiRCl \cdot AlR_2Cl$ (70).

5) ION REACTIVITY (SOLVENT INFLUENCE).

In the investigation of the medium polarity effect (i.e., dielectric constant) on the reaction rate of active ions with olefins(51), a linear dependence of $\log(K_{eff}/\eta)$ on $1/D$ was observed in five solvents and their mixtures (heptane & dichloroethane) (Fig. 3), where K_{eff} is an effective reaction rate constant of the complex $(C_5H_5)_2TiCH_3Cl \cdot AlCH_3Cl_2$ with an olefin, κ is specific electric conductivity, η is viscosity, and D is a dielectric constant. The value of $\log(K_{eff}/\eta)$ is proportional to the rate constant of the reaction of an active ion with an olefin. Thus the linear dependance corresponds to the well-known theoretical equation of an ion reacting with a neutral molecule:

$$\frac{d\ln k}{d(1/D)} = \frac{e^2 z_A^2}{2kT}\left(\frac{1}{r_A} - \frac{1}{r^*}\right) \qquad [20]$$

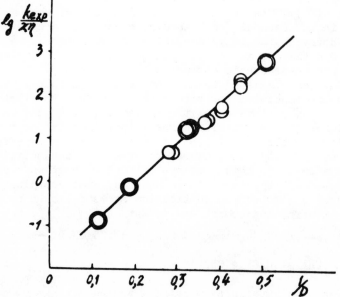

Fig. 3. Relation of $\log(k_{exp}/\kappa\eta)$ to $1/D$ for decene reaction with complex $(\pi\text{-}C_5H_5)_2Ti(CH_3)Cl \cdot Al(CH_3)Cl_2$.

where e is the electronic charge, Z_A is an ionic charge, and r_A, r^* are the radii of the ion and an activated complex, respectively. The values of rate constants of ionic interaction with decene and equilibrium dissociation constant of $(C_5H_5)_2TiCH_3Cl \cdot AlCH_3Cl_2$ have been reported(51), Table 2.

TABLE 2.

Ion $(C_5H_5)_2TiR^+$ *interaction constants with decene (k) and complex* $(C_5H_5)_2TiRClAlRCl_2$ *dissociation constant in Various solvents.*

	K	ΔQ, cal/mole	k, l/mole/sec	E cal/mole
Benzene + heptane	$7.8 \cdot 10^{-17}$	8,200	$26 \cdot 10^3$	4,500
Benzene	$1.3 \cdot 10^{-14}$	6,100	$5 \cdot 10^3$	5,600
Dichloroethane + heptane	$1.4 \cdot 10^{-11}$	-1,700	68	9,200
Dichloroethane	$9.4 \cdot 10^{-7}$	-2,200	9.3	10,300

As was seen from the works of Chirkov and his colleagues (7,54), who investigated a soluble catalytic system based on $(C_5H_5)_2TiCl_2$, an increase in the initiation rate was not the only effect produced by polar solvents containing chlorine. Thus, the study of the $(C_5H_5)_2TiCl_2$ + $AlEt_2Cl$ catalytic system in methylene chloride by EPR pointed at the existence of the following reactions(55):

$$(C_5H_5)_2Ti(IV)RCl \cdot AlRCl_2 \xrightarrow[AlR_2Cl]{k_1} (C_5H_5)_2Ti(III)Cl_2 \cdot AlRCl \xrightarrow[CH_2Cl_2]{k_2}$$

$$(C_5H_5)Ti(IV)Cl_3 \cdot AlRCl_2 \xrightarrow[AlR_2Cl]{} (C_5H_5)Ti(III)Cl_2 \cdot 2AlRCl_2 \quad [21]$$

In this system it is necessary also to take into account the interaction of low-valence transition metal derivatives with metallo-organic complexes (e.g. $(C_5H_5)_2$-TiCl, AlR_2Cl) with a chlorine-containing solvent.

It is evident that the solvent effect is rather complicated which includes both purely electrostatic action and specific chemical reactions.

6) *INTERACTION OF TRANSITION METAL ORGANIC DERIVATIVES WITH LEWIS BASES.*

It has been established by many experiments that the transition element-carbon σ- and π-bonds play an important role in the processes initiated by homogeneous complex catalysts. A number of works(56-66) were devoted to study factors effecting the stability of transition element-carbon bond and activation of an alkyl group in the coordination sphere of a transition metal. It is important to know why the reaction of olefin insertion in the M-C bond proceeds with high rate, while the dissociation of Ti-C or V-C bond does not occur. It seems likely that the high rate of insertion reaction is related to the fact that an olefin entering the coordination sphere of a complex decreases the dissociation energy of the transition element-carbon bond at the expense of its donor-acceptor properties. This hypothesis was confirmed by a number of studies(57,58,60,65) of the interaction between organic bases and olefins with organic transition metal derivatives.

The effect of organic base on the decomposition of cyclo-pentadienyl vanadium oxydichloride with a rupture of the $(C_5H_5)_2$-V bond has been reported(56). The kinetics and mechanism of this interaction was investigated using acetonitrile, dibutyl ether, and triphenyl phosphine as bases(57). The results of the kinetic study are given in Table 3.

TABLE 3.

Kinetic characteristics of reaction of $(C_5H_5)VOCl_2$ with organic bases. $[C_5H_5]VOCl_2 = 1.85 \cdot 10^{-3}$ *mole/l.*

Base (D)	E, kcal/mole	k, 1/mole/sec
Tetrahydrofuran	11.9	$4.9 \cdot 10^3$
Dibutyl ether	17.7	$4.4 \cdot 10^6$
Acetonitrile	15.4	$2.6 \cdot 10^5$
Triphenylphosphine	10.2	$1.8 \cdot 10^5$

The experimental data showed the rate determining step to be the interaction of a base with $(C_5H_5)VOCl_2$ with a simultaneous rupture of the $V-C_5H_5$ bond:

$$(C_5H_5)VOCl_2 + D \longrightarrow (C_5H_5) + VOCl_2 \cdot D \qquad [22]$$

The rate of disappearance of $(C_5H_5)-VOCl_2$ is the same as the rate of formation of $VOCl_2$.

The explanation of radical decomposition of $(C_5H_5)VOCl_2$ under the influence of solvating additives is the following: the energy of specific solvation by a donor is considerably larger for $VOCl_2$ than for $(C_5H_5)VOCl_2$. Actually, after the loss of a cyclopentadienyl radical, the low-level orbitals of vanadium are freed, one of which may be used to form a rather stable bond with a donor. That means that the C_5H_5-V bond energy is reduced in the presence of donor molecules.

Radical decomposition of CH_3TiCl_3 in the presence of organic bases was reported in the papers(58,59). The reaction of CH_3TiCl_3 with tetrahydrofuran in perdeuterotoluene produces monodeuterated methane indicating the formation of free methyl radicals in these systems(58). The formation of monodeuterated ethane was observed also in the reaction of CH_3TiCl_3 with hexene. The energy decrease of the Ti-C bond in CH_3TiCl_3 under the influence of organic bases was confirmed by i.r.-investigations(60). The formation of the $CH_3TiCl_3 \cdot D$ complexes resulted in a decrease of the Ti-C bond frequency.

In the case of reactions of organic bases with L_3CoCH_3, the behavior was different(61). If tetrahydrofuran and ethers were used as organic bases, there is practically no decomposition, i.e., the Co-C bond remained stable. The interaction of L_3CoCH_3 with unsaturated compounds led to various reactions of the methyl group in the coordination sphere ($L = P(Ph)_3$):

with hexene:

$$[23]$$

with butadiene:

$$[24]$$

with ethylene:

$$Ph_2PCoCH_3 \overset{(PPh_3)_2}{|} + 2CH_2=CH_2 \longrightarrow Co(C_2H_4)_2(PPh_3) \qquad [25]$$

The above results provide an explanation of high insertion rates of olefins in the M–C bond in coordination catalysts. An olefin or any other base tends to decrease the
transition element–carbon bond dissociation energy and
weakens its donor properties. The dative interaction of
metal d–electrons with the π–orbits of olefin molecules is
known to be the cause of formation of stable olefin complexes
with a number of metal compounds.

An olefin molecule inserts into the M–C bond without
radical decomposition of this bond. The formation of the
C–C bond and the new M–C bond occurs simultaneously with
scission of the M–C bond. The olefin coordination with transition metals considerably facilitates the process.

It is interesting to compare the reactivity of donor
molecules to $(C_5H_5)VOCl_2$, CH_3TiCl_3, and CH_3CoL_3. In case of
interaction with the Co–C bond, the following reaction order
has been observed:

$$C_4H_6 > C_2H_4 > C_6H_{12} >> C_6H_5C\equiv CH > C_2(CH_3)_4 >> C_4H_8O$$

One can see that tetrahydrofuran is very active in reactions
with high–valence compounds $(C_5H_5)VOCl_2$ and CH_3TiCl_3, but is
only weakly reactive with low–valence cobalt. This change in
the order of reactivity is consistent with increasing dative
$d \to \pi$ interaction with the decrease of number of d–electrons.

The change of the M–C bond under the influence of an
olefin has been discussed by Cossee(65). The coordination
of an olefin with CH_3TiCl_3 is thought to result in the formation of an octahedron with a new molecular orbital derived
from the metal d–orbital and π–antibonding orbital of ethylene.
This orbital is lower in energy than the usual metal 3d–levels.
Therefore, the electron from the metal–carbon bond can migrate
to this newly–formed orbital, thus weakening the Ti–C bond.

7) *MODIFICATION OF HOMOGENEOUS CATALYTIC SYSTEMS OF
OLEFIN POLYMERIZATION.*

The modification of complex catalysts widely reported in
the patent literature is in most cases arrived empirically

and not derived from an understanding of catalytic polymerization mechanism. We may mention here only two points of view derived from Shilov's and Chirkov's works. According to Chirkov's works the addition of small quantities of chlorine-containing compounds with a mobile chlorine atom can result in activating homogeneous catalytic systems in hydrocarbon solvents. This conclusion was proved correct(67). Thus, the addition of triphenylmethyl chloride to catalysts based on $(C_5H_5)_2TiCl_2$ increased the polymerization rate and its stability several times both in hydrocarbon solvents and in ethyl chloride. The vanadium catalytic systems in hydrocarbon solvents are stabilized by hexachlorpentadiene. However, the presence of chloride in these systems is undesirable technically. Besides, one alkyl group is lost in each of the successive reductions:

$$M-R \xrightarrow{\text{AlR}_2\text{Cl}} M$$

$$M \xrightarrow{\text{RCl}} MCl \qquad\qquad [26]$$

$$MCl \xrightarrow{\text{AlR}_2\text{Cl}} M-R$$

More significant is the oxidation of low valence transition metals derivatives by organic unsaturated compounds, for example:

$$Ti + CH_2=CH_2 + Ti \longrightarrow TiCH_2 - CH_2Ti \qquad [27]$$

In Dzhabiev's works these reactions were investigated in the $Ti(OR)_4 + AlR_3$ catalytic system(43) and elsewhere(68) for the $(C_5H_5)_2TiCl_2 + MgRHal$ catalytic system.

New M-C bonds formed in oxidative reactions can take part in the chain propagation reactions. Under such conditions transition metal alkylation occurs simultaneously with transition metal valence regeneration, i.e., in the oxidizing-reducing reactions an alkyl group is not lost and the system can be said to be truely catalytic.

Another way to modify homogeneous catalytic systems is associated with the ionic nature of the active centers. Addition to the system of compounds leading to greater polarization of a catalytic complex, i.e., an increase of the number of active ions (centers), resulted in an increase of the polymerization rate. For this purpose, Lewis acids were chosen. Thus $AlRCl_2$ was found to increase the rate of ethylene polymerization by the catalytic system $(C_5H_5)TiCl_3$ + $AlR_2Cl(69)$. The addition of $AlCl_3$ and $SnCl_4$ also increased ethylene polymerization rates.

217

A marked effect of increasing the transition metal activity is observed in supported catalysts. In these cases a considerable stabilization of the catalytic activity resulted from slow reduction of transition metal (Fig. 4).

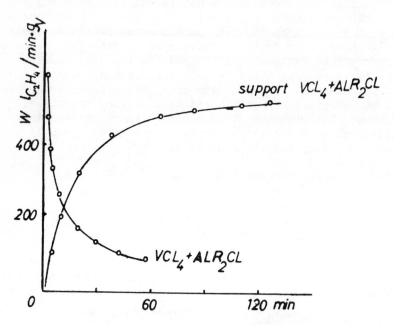

Fig. 4. Catalytic activity change of the VCl_4 + AlR_2Cl(I) and supported VCl_4 + AlR_2Cl systems.

In conclusion, the active center in homogeneous catalysis of olefin polymerization processes is the M-R group, where M is a transition metal (Ti, V, Zn, etc), and R is an alkyl radical, $PhCH_2$, Ph, π-alkyl group, etc. The structure of the R-group, the nature and energy of its bond with a transition metal determine the catalytic activity in many respects. The purpose of metallo-organic compounds (AlR_2Cl, MgRHal, ZnR_2, etc), is essentially to form and stabilize to some extent the active centers.

At present works have been started on synthesis of transition metal organic derivatives as models for the active centers of complex catalysts and measurement of catalytic activity. For example, the catalytic activity of $(C_5H_5)_2$-TiRCl in the presence of various aluminium-organic compounds is being studied in detail(70). The catalytic activity of tetrabenzyl titanium and its derivatives in diene and olefin polymerization without aluminium-organic compounds is determined(71). Polymerization with $(PhCH_2)_4Ti$ proceeds by inserting a monomer in the Ti-C bond(72) as in usual complex catalysts. These works are undoubtedly of great interest for the further development of polymerization catalysts.

LITERATURE CITED

(1) G. Natta, P. Pino, G. Mazzanti, U. Guannini, E. Mantica, and M. Perado, *J. Polym. Sci.* *26*, 120 (1957).

(2) D.S. Breslow, and N.R. Newburg, *J. Amer. Chem. Soc.* *79*, 5672, (1957).

(3) C.E.H. Bown, and R. Symocox, *J. Polym. Sci.* *34*, 139 (1959).

(4) G. Henrici-Olive, and S. Olive, *Adv. Polym. Sci.* *6*, 421 (1969).

(5) G. Henrici-Olive, and S. Olive, *Angew. Chem.* *84*, 4, 121 (1971).

(6) E.L. Fushman, V.I. Tsvetkova, and N.M.Chirkov, Isv. AN SSSR Chemistry, NII, 2075 (1965).

(7) E.L. Fushman, V.I. Tsvetkova, N.M. Chirkov, *Dokl. AN SSSR 164*, 1085 (1965).

(8) J.C.W. Chien, *J. Amer. Chem. Soc.* *81*, 86 (1959).

(9) J.C.W. Chien, and C.R. Boss, *J. Amer. Chem. Soc.* *83*, 3767 (1961).

(10) A.K. Zefirova, and A.E. Chilov, *Dokl. AN SSSR 136*, 599 (1961).

(11) A.E. Shilov, A.K. Shilova, and B.N. Bobkov, *Vis. Mol. Soed.* *4*, 1688 (1962).

(12) L.N. Stepovik, A.K. Shilova, and A.E. Shilov, *Dokl. AN SSSR 148*, 122 (1963).

(13) F.J. Karol, and W.L. Carrick, *J. Amer. Chem. Soc.* *83*, 2654 (1961).

(14) H. Bestian, and K. Clauss, *Angew. Chem.* *75*, 22 (1963).

(15) W.P. Long, and D.S. Breslow, *J. Amer. Chem. Soc.* *82*, 1953 (1960).

(16) H.J.M. Bartelink, H. Bos, and J. Smidt, Congres Ampere, Leipzig 13-16 Sept., (1961).

(17) T.S. Dzhabiev, F.S. Dyachkovskii, and A.E. Shilov, *Vis. Mol. Soed. A-XIII*, II, 2474 (1971).

(18) E.I. Nevelskii, F.S. Dyachkovskii, *Vis. Mol. Soed.*, II, 797 (1969).

(19) F.S. Dyachkovskii, *Vis. Mol. Soed.* *7*, 114 (1965).

(20) E.A. Grigorian, F.S. Dyachkovskii, G.M. Khvostik, A.E. Shilov, *Vis. Mol. Soed. A-9*, 1233 (1967).

(21) H. Bestian, and K. Clauss, *Angew. Chem., Intern. Ed. 2*, 704 (1963).

(22) F.S. Dyachkovskii, M.L. Eritsian, O.E. Kashireninov, B. Matyska, K. Mach, M. Svestka, and A.E. Shilov, *Vis. Mol. Soed., A-II*, N3, 543 (1969).

(23) R.D. Bushik, and R.S. Stearns, *J. Polym. Sci. A1*, 215 (1966).

(24) J.V. Nicolescu, and E.M. Angelescu, *J. Polym. Sci. A3*, 1227 (1965).

(25) A.W. Langer, *J. Macromol. Sci. Chem. A-4*, 775 (1970).

(26) A.P. Khrushch, and A.E. Shilov, *Kinetika i Kataliz 11*, 1, 86 (1970).

(27) A.E. Shilov, A.K. Zefirova, and N.N. Tikhomirova, *J. Fis. Khim. 33*, 2113 (1959).

(28) A.K. Zefirova, N.N. Tikhomirova, and A.E. Shilov, *Dokl. AN SSSR 132*, 1082 (1960).

(29) T.S. Dzhabiev, and A.E. Shilov, *Zh. Strukt. Khim. 6*, 302 (1965).

(30) S.O. Shulindin, N.N. Tikhomirova, A.E. Shilov, and A.K. Shilova, *Zh. Strukt. Khim. 2*, 740 (1961).

(31) G.A. Abakulov, A.E. Shilov, and S.V. Shulindin, *Kinetika i Kataliz 5*, 228 (1964).

(32) A.E. Shilov, and N.P. Bubnov, *Izv. AN SSSR, khim. N3*, 381 (1958).

(33) T.S. Dzhabiev, R.D. Sabirova, and A.E. Shilov, *Kinetika i Kataliz 5*, 411 (1964).

(34) B.A. Ierusalimskii, Van Fon-Sun, and A.P. Kavunenko, *Mezhd. Simpos. Makromol. Khim., Moskva, sekts. II*, 355 (1960).

(35) Van Fon-Sun, B.A. Dolgoplosk, and B.L. Ierusalimskii, *Izv. AN SSSR khim. N3*, 469 (1960).

(36) V.A. Kropachev, B.A. Dolgoplosk, N.M. Geller, and M.N. Zelenina, *Izv. AN SSSR khim. N6*, 1044 (1965).

(37) W.L. Carrick, W.T. Reichle, F. Pennella, and J.J. Smith, *J. Amer. Chem., Soc. 82*, 3887 (1960).

(38) F.S. Dyachkovskii, and N.E. Khrushch, *Zh. Ob. Khim. 41*, 8, 1779 (1971).

(39) H.D. Vries, *Recueil. trav. chim. 80*, 866 (1961).

(40) F.A. Cotton, *Chem. Revs. 55*, 551 (1955).

(41) Zh. "Sekubai", *9* 1-41 (1967).

(42) L.S. Bresler, I.I. Poddubny, T.K. Smirnova, A.S. Khachaturov, and I.Y. Tsereteli, *Dokl. AN SSSR 210*, 847 (1973).

(43) Kandid. Dissert., T.S. Dzhabiev, Moskva, 1971.

(44) G.D. Bukatov, V.A. Zakharov, Y.I. Ermakov, *Kinetika i Kataliz 12*, 505 (1971).

(45) J. Boor, *Macromol. Revs. 2*, 170 (1967).

(46) F.X. Werler, C.J. Benning, W.R. Wszolek, G.E. Ashly, *J. Polym. Sci. A-1, 6*, 743 (1968).

(47) A.E. Shilov, A.K. Shilova, B.N. Bobkov, *Visok. Molek. Soed. 4*, 1688 (1962).

(48) L.P. Stepovik, A.K. Shilova, and A.E. Shilov, *Dokl, AN SSSR 148*, 122 (1963).

(49) F.S. Dyachkovskii, P.A. Yarovitskii, V.F. Bistrov, *Vis. Molek. Soed.* *6*, 659 (1964).

(50) F.S. Dyachkovskii, A.K. Shilova, and A.E. Shilov, *J. Polym. Sci.* *C16*, 2333 (1967).

(51) O.N. Babkina, E.A. Grigorian, F.S. Dyachkovskii, A.E. Shilov, and N.M. Shuvalova, *Zh. Fiz. Khim.* *43*, 7 (1969).

(52) T.S. Dzhabiev, F.S. Dyachkovskii, I.I. Shishkina, *Izv. AN SSSR, khim.* *N6*, 1238 (1973).

(53) T.S. Dzhabiev, F.S. Dyachkovskii, and N.D. Karpova, *Kinetika i Kataliz* *15*, 67-71 (1974).

(54) E.L. Fushman. M.P. Gerasina, S.P. Utkin, A.A. Brikenshtein, V.I. Tsvetkova, and N.M. Chirkov, *Plast. Mass.* *N10*, 196.

(55) E.A. Grigorian, N.M. Semenova, and F.S. Dyachkovskii, *Izv. AN SSSR, khim.* *N12*, 2719 (1972).

(56) S.V. Shulindin, Kandid. Dissert., Moskva, 1966.

(57) F.S. Dyachkovskii, N.E. Khrushch, and A.E. Shilov, *Kinetika i Kataliz* *8*, 1230 (1967).

(58) F.S. Dyachkovskii, N.E. Khrushch, and A.E. Shilov, *Kinetika i Kataliz* *9*, 1006 (1968).

(59) K.H. Thiele, and W.Grahlert, *Weissen. Z. der Techn.* *13*, Hf2, (1969).

(60) O.S. Roshchupkina, N.E. Khrushch, F.S. Dyachkovskii, and Y.G. Borodko, *Zh. Fiz. Khim.* *46*, 1329 (1971).

(61) N.E. Khrushch, F.S. Dyachkovskii, and A.E. Shilov, *Zh. Ob. Khim.* *11*, 1726 (1970).

(62) A.N. Plusnin, B.A. Uvarov, V.I. Tsvetkova, and N.M. Chirkov, *Izv. AN SSSR khim.* *N10*, 2324 (1969).

(63) R.J.H. Clark, M. Coles, and A.J. McAlees, *V Mezhd. Kongress po Metallorg. Khim., Moskva* t. *II*, 626 (1971).

(64) R.J.H. Clark, M. Coles, *J. Chem. Soc. Dalt. tr., N22*, 2454 (1972).

(65) P. Cossee, *J. Catal.* *8*, 80 (1964).

(66) P.E. Matkovskii, L.I. Chernaja, F.S. Dyachkovskii, and N.M. Chirkov, *Zh. Ob. Khim.* *N3*, (1974).

(67) P.E. Maykovskii, T.S. Dzhabiev, F.S. Dyachkovskii, G.A. Beihold, A.A. Brikenshtein, and N.M. Chirkov, *Vis. Molek. Soed. A-9*, 1762 (1971).

(68) Y.G. Borodko, E.F. Kyashina, V.B. Panov, and A.E. Shilov, *Kinetika i Kataliz* *14*, 255 (1973).

(69) P.E. Matkovskii, G.A. Beihold, L.N. Russian, A.A. Brikenshtein, and N.M. Chirkov, *Vis. Molek. Soed. A-16*, 176 (1971).

(70) L.F. Borisova, E.L. Fushman, E.J. Vizen, N.M. Chirkov, *Europ. Polym. J.* *9*, 953 (1973).

(71) C.J. Attridgr, and R. Jackson, *J. Chem. Soc. Chem. Commun.* *4*, 132 (1973).

(72) D.G.H. Ballard, N. Heap, B.T. Kilbourn, and R.J. Wyatt, *Makromol. Chem.* *170*, 1 (1973).

Transition Metal Alkyl Polymerization Catalysts

D.G.H. BALLARD

Corporate Laboratory
Imperial Chemical Industries Ltd.
The Heath, Runcorn, Cheshire, England

1) *INTRODUCTION*

The significance of the discovery of Professor Ziegler and his colleagues at the Max Plank Institute at Mulheim of transition metal polymerization catalysis is readily obvious to industrial scientists. Like others here, I am of the opinion that the discovery contained within it a vitally important part of chemistry which if we could understand it would lead on to important inventions. We at the Corporate Laboratory of ICI have set ourselves the task of trying to unlock Pandora's Box and as you will see from this lecture, we have not altogether been unrewarded.

Our approach to the problem of understanding the mechanism of action of polymerization catalysts of the Ziegler type has been to synthesize models of known structure and to study their polymerization behaviour using kinetic, spectroscopic, and radiochemical techniques. We recognized at the beginning that there was considerable disagreement about the structure of the catalytic center mainly because of the uncertain role of the aluminium alkyl. One school suggested that the aluminium alkyl was exclusively responsible for activating the metal by forming a transition metal alkyl bond giving an intermediate of the type:

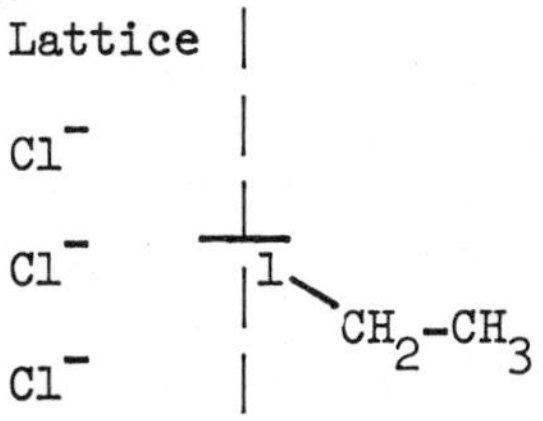

Lattice |

Cl^-

Cl^-

Cl^-

$CH_2\text{-}CH_3$

and had no direct role in the growth reactions. An alternative set of ideas suggested that bridged complexes were formed, for example:

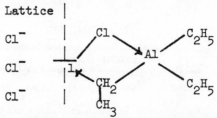

and growth reaction again involved exclusively the transition metal. Compounds of the first type are transition metal alkyl compounds and since these could be made we decided to study this class of substances. In doing so, we have developed a new family of catalysts for polymerization and other purposes. The bridged complexes however present formidably synthetic problems and as yet little headway has been made on the synthetic active catalysts based on this structural form. Finally, we have also looked at organometallic catalytic reactions as a class and shown that there are considerable similarities between them. Moreover, if one considers them as a group of organic reactions and forgets about the metal, one can arrive at simple and comprehensible mechanisms for describing this chemistry(1).

2) *CATALYTIC ACTIVITY AND STRUCTURE OF TRANSITION METAL ALKYLS*

Transition metal alkyl compounds are correctly divided into two groups(2), π-complexes and sigma complexes. The former include $U(\pi\text{-}C_8H_8)_2$, $Cr(\pi\text{-}C_6H_6)_2$ and $Ti(\pi\text{-}C_5H_5)_2$, $Zr(C_3H_3)_4$, etc. In the π-allyl compounds for example, the bonding electrons are delocalized over all the carbon atoms in the ring giving a symmetrical molecule in which the metal-carbon bond distances are identical(3,1). It has been found (2,4) that only the π-allyl compounds are active catalysts for the polymerization of olefins

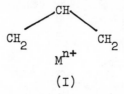

(I)

and vinyl monomers. For example $Zr(\text{allyl})_4$ and $Cr(\text{allyl})_3$ will polymerize ethylene with the activities given in Table 1,

TABLE 1.

Polymerization of ethylene in the dark with transition metal alkyls.[a]

Catalyst	Activity α gms/mM/atm./hr.
$Zr(allyl)_4$	2.00
$Cr(allyl)_3$	0.3
$Zr(allyl)_2(Cpd)_2$	Nil
$(allyl)_2Cr-Cr(allyl)_2$	Nil
$[Cpd]_2Zr[CH_2Si(C_6H_5)_3]Cl$	Nil
$Zr[CH_2 Si(CH_3)_3]_4$	0.9
$Ti[CH_2Si(CH_3)_3]_4$	0.2
$Ti[C_6H_5CHSi(CH_3)_2]_4$	0.2
$Zr[CH_2 C(CH_3)_3]_4$	0.8
$[Cpd]_2Ti[CH_2 Si(CH_3)_3]_2$	Nil
$Ti(benzyl)_4$	0.2
$Zr(benzyl)_4$	0.8
$Hf(benzyl)_4$	0.42
$Zr[CH_2 S CH_3]_4$	0.1
$Zr[CH_2 OCH_3]_4$	0.5

[a]In toluene at 80°C, $P_{C_2H_4}$ = 10 atm., [Catalyst] = 3 mM (2).

but the mixed π-complex $(Cpd)_2Zr(allyl)_2$ is inactive. Also compound with metal-metal bonds are not polymerization catalysts.

Substitution of the allyl group with halogen atoms can increase activity markedly and the species $Zr(allyl)Br_3$ is probably the most active of the simply allyl compounds. For short periods, activities of 800 gm mM/atm/hr have been obtained at 80°C in toluene with ethylene, giving terminally insaturated waxes where $\overline{M}_n \approx 800$. The majority of the monomers are polymerized by Br_3ZrH generated *in situ* in accordance with the following mechanism(2).

$$Br_3Zr\overset{H_2C}{\underset{\underset{H_2}{C}}{\diagdown}}CH + CH_2{=}CH_2 \rightleftharpoons Br_3Zr{\cdot}CH_2{\cdot}CH{=}CH_2 \qquad [1]$$

$$\qquad [2]$$

$$Br_3Zr(CH_2CH_2)_{n+1}R \xleftarrow{n(CH_2=CH_2)} Br_3Zr(CH_2{\cdot}CH_2)R \qquad [3]$$

$$\text{(III)} \qquad\qquad\qquad \text{(II)}$$

$$\text{(III)}\longrightarrow Br_3ZrH + CH_2{=}CH(CH_2{\cdot}CH_2)_nR \qquad [4]$$

$$Br_3ZrH \xrightarrow{n(CH_2=CH_2)} \text{(III)}(R = C_2H_5) \qquad [5]$$

The propagating species at any instant of time probably has the structure (III), the actual monomer insertion process being identical to processes 1 and 2. Competing with the propagation reaction is removal of a hydrogen atom from the β-carbon atom by a mechanism which we will all discuss later giving the hydride Br_3ZrH and a terminally unsaturated polymer. Process 5 is preceded by a very rapid realkylation of the latter with monomer. This is the closest analogy we have to the concept of Ziegler center described earlier.

From the point of view of polymerization catalysis the largest group is the sigma complexes(2), which can be represented by $M^{n+}(R_1R_2CY)_n$ where R_1 and R_2 are hydrogen or carbon atoms and Y can be $M'(CH_3)_n$; M' is Si, Ge or Sn or aryl, OCH_3, SCH_3; or R_1, R_2 and Y can form part of a bicyclic ring system (with no β-hydrogen atoms) such as the norbornyl(5) group. Some examples are $Zr(benzyl)_4$, $Ti(CH_2Si(CH_3)_3)_4$, $Cr(norbornyl)_4$ etc. Nearly all of the sigma complexes mentioned can be used as polymerization catalysts (Table 1) and like the transition metal alkyls have only weak activity for the polymerization of olefins. An important difference between the transition metal allyl compounds and benzyl compounds is that the latter will polymerize styrene whereas the former does not.

It has been demonstrated(2) that more active catalysts can be obtained from transition metal alkyls if interaction of metal centers can be prevented. This requirement can be met by using the strongly acidic -OH groups on the surface of alumina and silica. These, on reaction with transition metal alkyls were found to give highly active polymerization catalysts of long lifetime(20,22).

The surface of silica and alumina freed from physically adsorbed water contain acidic OH groups which will react with transition metal alkyls. These reactions can be followed in the infrared(2). It has been found that transition metal alkyl compounds can react with the OH groups in more than one way and the product obtained depends on several factors. For example, $Zr(allyl)_4$ reacts with silica pre-dried at 200°C to give two molecules of propene per metal atom utilizing in the course of this process two OH groups per metal atom. The chemistry of the process is accurately described by the equation:

$$\begin{array}{c}
\equiv Si-OH \\
O \\
\equiv Si-OH
\end{array}
+ Zr(allyl)_4 \longrightarrow
\begin{array}{c}
\equiv Si-O \\
O \qquad Zr \\
\equiv Si-O
\end{array}
\begin{array}{c}
C_3H_5 \\
\\
C_3H_5
\end{array}
+ 2C_3H_6 \quad [7]$$

The structure for the transition metal center is confirmed by measuring the amount of propene produced on reaction of (X) with n-butanol;

$$\begin{array}{c}
\equiv Si-O \\
O \qquad Zr \\
\equiv Si-O
\end{array}
\begin{array}{c}
C_3H_5 \\
\\
C_3H_5
\end{array}
+ 2 \text{ BuOH} \longrightarrow
\begin{array}{c}
\equiv Si-O \\
O \qquad Zr \\
\equiv Si-O
\end{array}
\begin{array}{c}
OBu \\
\\
OBu
\end{array}
+ 2C_3H_6 \quad [8]$$

and by infrared study(2). Similar observations have been made with $Zr(CH_2C_6H_5)_4$, $Zr(CH_2Si(CH_3)_3)_4$, $Zr(CH_2OCH_3)_4$, etc. Silica can be replaced by alumina and other matrices, giving transition metal centers with structures related to (X) in which the organic ligands are $CH_2C_6H_5$, $CH_2Si(CH_3)_3$, etc., etc.

If some of the hydrocarbyl ligands on Ti, Zr or Hf alkyls are replaced by halogen atoms, transition metal centers of type (XI) and (XII) are obtained:

(X)   (XI)   (XII)   (XIII)

227

Zr(allyl)$_3$Cl gives primarily but not exclusively a center of type (XI) since only two molecules of propene per Zr atom are evolved on reaction with silica, and halogen compounds are not found in the solvent or in the gas. The compound SiO$_2$/Zr-(allyl)$_3$Cl gives one molecule of propene per zirconium atom with excess butanol, and one molecule of HCl per zirconium atom on reaction with excess benzoic acid solution. The structure of (XII) was determined in a similar manner. Chromium allyl gives a transition metal center with structure (XII).

It is evident from the structures of the transition metal centers (X), (XI) and (XIII) that the distribution of –OH groups on the surface is not random but that a significant number of –OH groups occur in pairs. This implies that on 200°C dried silica the transition metal centers are approximately 10 Å apart.

It is also possible to remove the alkyl groups completely from species of the type (X) by hydrogenolysis giving metal hydride compounds which are also active polymerization catalysts(10).

$$\text{(XIV)} \qquad\qquad [9]$$

Alternatively, the borohydrides of transition metals(11) may be used

$$\text{(XV)} \qquad\qquad [10]$$

in these compounds the bonding to the metal is through the H-atom and not the boron. During the polymerization with these catalysts B$_2$H$_6$ is evolved initially and they behave as transition metal halides.

3) STUDIES OF THE EQUILIBRIUM BETWEEN METAL ALKYL COMPOUNDS AND OLEFINS.

Despite the fact nearly every mechanism for the polymerization of olefins using transition metal compounds of Ti, Zr and Hf write the initial reactions showing co-ordination of the olefin to the transition metal, as for example in eq. 1,

no direct evidence for the existence of such complexes has
been obtained. It is not acceptable to infer the existence
of olefin complexes from the known compounds of platinum and
palladium(7). The olefin compounds derived from the latter
metals depend for their stability on d-electrons which are
essential to the bonding of the olefin; Ti, Zr, Hf in the
four valent state have no d-electrons.

Indirect evidence has been obtained for the existence
of species (XVI) from studies of the interaction of $Zr(benzyl)_4$
with Lewis bases.

$$(C_6H_5CH_2)_4Zr + CH_2=CHR \longrightarrow (C_6H_5CH_2)_4Zr \leftarrow \begin{matrix} CH_2 \\ \| \\ CHR \end{matrix} \qquad [11]$$

$$(XVI)$$

It has been shown that $Zr(benzyl)_4$ react reversibly with compounds such as pyridine, quinoline, tri-n-butyl phosphine
oxide, giving compounds such as $Zr(benzyl)_4 \cdot L$ and $Zr(benzyl)_4L_2$.
They are readily detected by observation of the methylene
protons of the benzyl groups or the relevant ligand protons
in the 220-MHz NMR spectrometer(12). For example, pyridine
reacts in the following way.

$$Zr(CH_2C_6H_5)_5 + C_5H_5N \rightleftharpoons \begin{matrix} & C_6H_5N & \\ C_6H_5CH_2 & \downarrow & CH_2C_6H_5 \\ & Zr & \\ C_6H_5CH_2 & & CH_2C_6H_5 \end{matrix} \qquad [12]$$

$$(XVII)$$

The spectroscopic studies show the coordination complex is in
equilibrium with free $Zr(benzyl)_4$ and pyridine but this equilibrium is predominantly to the right-hand side of eq. 12.
In fact both pyridine complexes have been isolated. In view
of these results it is natural to assume that since ethylene
may also be considered as Lewis bases that species of type
(XVI) will also be formed. However, it was found using
similar spectroscopic measurements to the above that such a
complex could not be detected (13). From this it was concluded that if the complex was formed its concentration was
small, probably less than one percent of the $Zr(benzyl)_4$
present. Kinetic studies to be discussed supported the
existence of such a complex, and indicated that its concentration was small but that it was formed very rapidly.

Values for the equilibrium constant for the complex
formation with a silanomethyl compound have recently been
measured(14) by modification to the well-known gas-liquid
chromatography method previously used to measure the equili-

brium constant for complexes of olefins with $AgNO_3$(15), $Rh(CO)_2$ (acac)(16), and $IrI(CO)(PPh_3)_2$(17). The values for the equilibrium constant confirm that the concentration of complex (V) is indeed very small with the polymerization conditions used, in fact, less than 0.1% of the catalyst present will be in this form. A value of $K = 12.6 \pm 0.6$ M^{-1} for the pyridine/ $Zr(benzyl)_4$ 1:1 adduct(20) has been obtained at 78°C, hence the values in Table 2 are reasonably consistent with the relative basicities of pyridine and ethylene.

TABLE 2.

Dependence of molecular weight $(\overline{M}_n)$ on initial $Zr(benzyl)_4$ concentrations for styrene polymerization.[a]

$[C]_0$, $M \cdot 10^2$	0.3	1.0	3.0	3.0	4.7
$\overline{M}_n \cdot 10^{-3}$	150	180	150	130	150

[a]In toluene at 30°C, initial monomer concentration is 5 M.

An interesting result in Table 2 is that butadiene coordinates less readily than ethylene, this agrees with previous observations with Ag^I salts(10), where the equilibrium constants were in the order, ethylene > propene > butene-1 > butadiene. Intuitively one would guess that butadiene would be more strongly coordinated to the transition metal.

4) POLYMERIZATION OF STYRENE BY ZIRCONIUM BENZYL & HOMOGENEOUS MODEL FOR THE STUDY OF THE MECHANISM OF POLYMERIZATION.

$Zr(benzyl)_4$ is a slow initiator for the polymerization of styrene. At low conversions (up to 2%) the conversion-time curve is almost linear and initial rate curves obtained using dilatometry give accurate and reproduceable data. Polymerization to much higher conversions can be followed gravimetrically to give a similar curve. A typical example is shown in Fig. 1. At these higher conversions the system is still completely homogeneous and no darkening of the reaction medium is observed. Observations with infra-red radiation show that at these high conversions there is still present a large concentration of substances containing benzyl groups attached to metal atoms.

The initial rates of polymerization of styrene (R) at 30°C in toluene for different initial concentrations of Zr(benzyl)$_4$ ([C]$_o$), whilst maintaining the initial monomer concentration ([M]$_o$) constant, is shown in Fig. 2. The relationship between the initial rate of polymerization of styrene

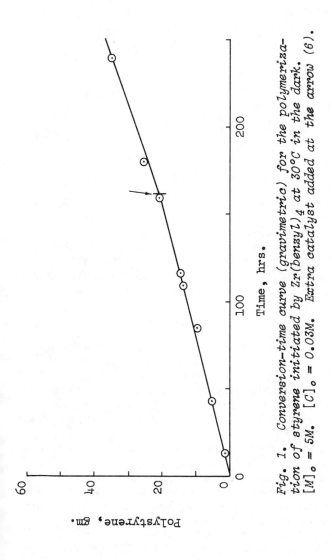

Fig. 1. Conversion-time curve (gravimetric) for the polymerization of styrene initiated by Zr(benzyl)$_4$ at 30°C in the dark. [M]$_o$ = 5M. [C]$_o$ = 0.03M. Extra catalyst added at the arrow (6).

and monomer concentration was complex, and a plot of $[M]_o/R$ against $1/[M]_o$ gave the best fit, this is shown clearly in Fig. 3.for $30^\circ C$ and $40^\circ C$. The results of the two sets of experiments can be summarized by the expression:

$$R = -d[M]/dt = (A[C]_o[M]_o^2)/(1 + B[M]_o) \qquad [13]$$

where A and B are constants for any given temperature.

The initial rate of polymerization of methyl methacrylate initiated by chromium allyls (21) in toluene showed identical dependences on monomer and catalyst concentrations, as Zr-(benzyl)$_4$ initiated polymerization of styrene. Some data for the monomer dependence is shown in Fig. 4.

The number average degree of polymerization $\overline{P}_n$ of unfractionated polystyrenes obtained with Zr(benzyl)$_4$ in toluene as solvent changes with conversion in the manner shown in Fig. 5. Within the limits of experimental error there is very little change in $\overline{P}_n$ up to five percent conversion, though overall there appears to be a small increase in molecular weight at conversions above this. The molecular weights of the polystyrenes appears to be independent of the catalyst concentration over a sixteen-fold change in the ratio of $[M]_o/[C]_o$; this is seen in Table 2. It was found, however, that the molecular weight is markedly dependent on the monomer concentration, and a plot of the reciprocal of $\overline{P}_n$ against the reciprocal of the initial monomer concentration used in their preparation gives an excellent straight line (Fig. 6) indicating that $\overline{P}_n$ is related to the monomer concentration by the expression similar to

$$\overline{P}_n = A'[M]_o/(1 + B'[M]_o) \qquad [14]$$

The polymerization of methyl methacrylate initiated by chromium allyls gave identical results(14), although the molecular weights of the polymers obtained were lower.

The molecular weight distributions obtained for polymers prepared with these catalysts have a polydispersity close to 2, over a five-fold change in molecular weight. This is shown in Table 3.

It was shown(13,14) with transition metal alkyls containing C^{14} in the alkyl group that the polymers produced contain alkyl group covalently bonded to the polymer chain. On average each chain contained rather more than one alkyl group derived from the catalyst. The excess over one per chain has been interpreted as an error introduced by the presence of a low molecular weight tail in the distribution. It was also demonstrated that if each molecule of Zr(benzyl)$_4$ produced one polystyrene chain then only one hundredth of the initiator was employed at conversions to polystyrenes of about ten percent.

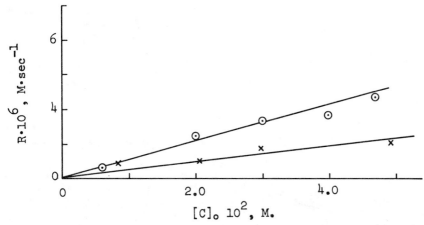

Fig. 2. *Dependence of the initial rate of polymerization in the dark on Zr(benzyl)$_4$ concentration, solvent toluene. (a) $[M]_o = 5.0$ M; (b) $[M]_o = 3.0$ M; Temperature 30°C (6).*

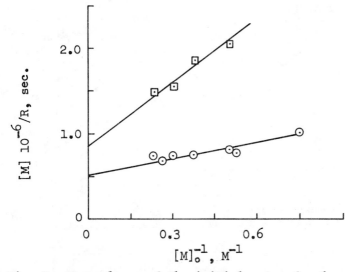

Fig. 3. *Dependence of the initial rate of polymerization of styrene initiated by Zr(benzyl)$_4$ in the dark on the concentration of styrene: $[C]_o = 0.03$ M, (a) measurements at 30°C; (b) measurements at 40°C, solvent toluene (6).*

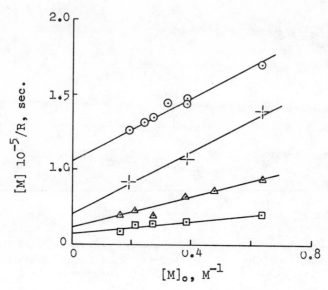

Fig. 4. *Dependence of the initial rate of polymerization of methyl methacrylate initiated by Cr(allyl)₃ in toluene a at different temperatures (14); (a) 0°C, [C]ₒ = 0.03 M; (b) 5°C, [C]ₒ = 0.033 M; (c) 30°C, [C]ₒ = 0.02 M; (d) 40°C, [C]ₒ = 0.02 M.*

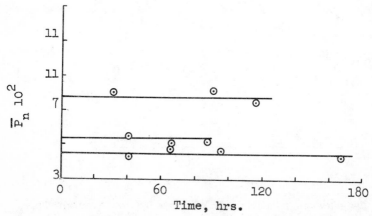

Fig. 5. *The number average degree of polymerization of polystyrenes at different conversions prepared in toluene using Zr(benzyl)₄ as initiator in the dark. [M]ₒ values: (a) 5.0 M; (b) 4.0 M; (c) 3.0 M. [C]ₒ = 0.03 M; temperature 30°C (6).*

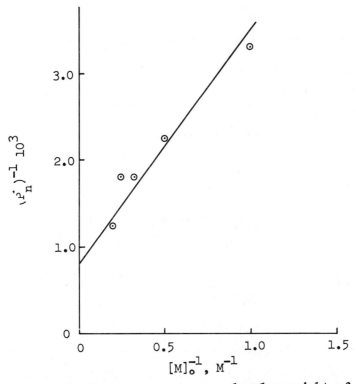

Fig. 6. Changes in the number average molecular weight of polystyrenes prepared in toluene at 30°C using Zr(benzyl)₄ as initiator at different initial monomer concentrations (6).

Using a C.A.T. with a 220 NMR spectrometer scanning over 200 times revealed an end-group of the type $C_6H_5CH=CH-$ and no end-groups similar to $-CH(C_6H_5)-CH_2C_6H_5$. The radiochemical data as well as the spectroscopic data suggest that the propagating center has structure (XVIII) and the polymer derived from it by β-hydrogen abstraction has structure (XIX). If the number of benzyl

$$(C_6H_5CH_2)_3ZrCH(C_6H_5)CH_2(CH(C_6H_5)CH_2)_{n-1}CH_2C_6H_5 \;\;(XVIII) \longrightarrow$$

$$(C_6H_5CH_2)_3ZrH+C_6H_5CH=CH_2(CH(C_6H_5)CH_2)_{n-1}CH_2C_6H_5 \;\;(XIX) \qquad [14]$$

TABLE 3.

Dependence of molecular weights and polydispersity on initial monomer concentrations.[a]

$[M]_o^b$	$\overline{M}_n$	$\overline{M}_w$	$\overline{M}_w/\overline{M}_n$
M			
1.0	30,000	---	---
2.0	44,100	83,500	1.89
3.0	55,400	121,000	2.32
4.0	55,100	125,000	2.27
5.0	82,000	202,000	2.47
6.5	158,000	358,000	2.26

[a]Styrene polymerization in toluene at 30°C catalyzed by $[Zr(benzyl)_4]_o = 0.03$ M. [b]Initial styrene concentration.

groups per chain had been less than unity this would have shown that the hydride $(C_6H_5CH_2)_3Zr-H$ realkylates with styrene to re-establish the propagating center. This certainly occurs in the case of ethylene but for reasons which are not understood this does not happen in the case of styrene, hence eq. 14 is a termination process and the hydride is inactive. Evidence was therefore sought for the existence of zirconium hydrides. Perdeuteromethanol reacts with metal hydrides giving deuterium hydride which can be measured in small quantities in the mass spectrometer. The reactions involved are probably as follows:

$$(C_6H_5CH_2)_3ZrH + CD_3OD \longrightarrow 3C_6H_5CH_2D + HD + Zr(OCD_3)_4 \qquad [15]$$

It follows that each molecule of HD formed corresponds to a molecule of metal hydride. Measurements of HD showed that one percent of metal hydride was present as an impurity in the $Zr(benzyl)_4$ solution in toluene catalyst. On adding styrene monomer the hydride did not disappear from the reaction mixture but progressively increased as the polymerization proceeded. It was estimated that if the hydride had the empirical formula $[(C_6H_5CH_2)_3ZrH]_n$, the amount formed corresponded to one molecule per chain. The persistence of this hydride in solution probably results from dimerization giving species of the type (XX).

$$(C_6H_5CH_2)_3Zr \underset{H}{\overset{H}{\diagup\diagdown}} Zr(CH_2C_6H_5)_3$$

(XX)

The number of polymer chains attached to zirconium atoms can be measured by treating the reaction mixture with excess tritium oxide giving a radioactive polymer (XXI). Radiochemical measurements can therefore be used to determine the number of polymer chains attached to metal atoms during the polymerization.

$$(C_6H_5CH_2)_3Zr(\underset{\underset{C_6H_5}{|}}{CH-CH_2})_nCH_2C_6H_5 + T_2O$$

[16]

$$\longrightarrow 3C_6H_5CH_2T + Zr(OT)_4 + T-(\underset{\underset{C_6H_5}{|}}{CH-CH_2})_nCH_2C_6H_5$$

(XXI)

In a typical run the final activity of the polymer, after vigorous purification, was 1200 counts/min/gm polymer, and this activity could not be reduced further by additional purification. This is therefore evidence that there are polymer chains directly attached to zirconium atoms. Since each gram-atom of tritium was initially equivalent to $9 \times 1.11 \times 10^8$ counts/min, the number of gram-atoms of tritium present in each gram polymer is $1200/(9 \times 1.11 \times 10^8) = 1.2 \times 10^{-6}$, the latter is also the number of polymer chains originally attached to zirconium atoms. From the molecular weight of the polymer ($\overline{M}_n = 130,000$) the total number of polymer chains in each gram of polymer is 7.7×10^{-6} moles. It follows therefore that:

$$\frac{\text{Number of polymer chains attached to Zr atoms}}{\text{Number of polymer chains}} = \frac{1}{6.4}$$

This confirms the observation derived from measurements of zirconium hydride concentrations and terminal double bonds that a majority of polymer chains become detached from the transition metal centers.

The total concentration of polymer chains, Q, attached to zirconium atoms was also determined as a function of conver-

237

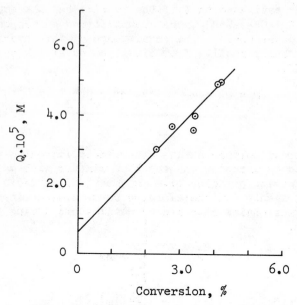

Fig. 7. The polymerization of styrene by Zr(benzyl)₄. Measurements of the total number of polymer chains Q, attached to zirconium atoms as a function of conversion. $[M]_o = 3.0$ M, $[C]_o = 0.03$ M. Temperature 30°C. Solvent toluene(6).

sion. A typical result obtained is shown in Fig. 7. It is evident from this data that Q increases in a linear way with conversion. It has been shown, however, that the rate of polymerization remains constant over this interval (Fig. 1). Q cannot therefore correspond to the number of propagating centers present. It follows that there are present in the reaction mixture chains attached to metal atoms which do not participate in the polymerization process. This is reasonably good evidence for a second termination process which does not involve detachment of the polymer chain from the metal atom. The nature of the chemical species produced by this termination process is not known, but an inactive metal polymer species could result from attack of a monomer on a second benzyl group on the metal atom:

$$(C_6H_5CH_2)_3Zr-P_n + CH_2=CHC_6H_5 \longrightarrow (C_6H_5CH_2)_2Zr \overset{\displaystyle \overset{C_6H_5}{\underset{|}{CH-CH_2CH_2C_6H_5}}}{\underset{\displaystyle P_n}{\diagdown}}$$

[17]

$$(C_6H_5CH_2)_2Zr \longrightarrow P_n$$

(XXII)

It has been shown(2) that transition metal alkyl compounds containing Cp and C_6H_5 groups, π-bonded to the metal inactivate the metal center for polymerization. It has also been shown by Aresta and Nyholm(18), in the platinum series, that 5 or 6 membered rings containing only sigma and π-carbon-to-metal bonds are very stable compounds. These observations add chemical plausibility to reaction 12.

The quantity Q therefore is the sum of two quantities, the number of polymer chains attached to metal atoms which are active in polymerization ($\sum_1^\infty [D_n]$) and which have the structure (VII), and species with structure (XI) which are inactive ($\sum_1^\infty [Z_n]$). In addition there are those chains with structure (XII) in which the monomer is coordinated to the metal [$\sum_1^\infty [E_n]$]; because of the equilibrium constant measurements, (section 2.2), these are considered to be negligibly small compared to $\sum_1^\infty [D_n]$. The quantity Q can therefore be represented by:

$$Q = \sum_1^\infty [D_n] + \sum_1^\infty [Z_n]$$

[18]

An analysis of the kinetics shows that $\sum_1^\infty [Z_n]$ will increase linearly with conversions at low conversions, whereas the quantity $\sum_1^\infty [D_n]$ will remain constant since the rate of polymerization is constant. It is therefore possible to obtain the important quantity $\sum_1^\infty [D_n]$, the stationary number of growing polymer chains from the intercept obtained in Fig. 7, by extrapolating values of Q to zero conversion.

The polymerization of styrene by Zr(benzyl)$_4$ has the characteristics of producing high molecular weight polymers by a very slow polymerization process. This can be reasonably explained by a slow initiation process followed by a fast propagation reaction. These two processes are probably chemically similar but differ significantly in rate for structural reasons.

239

Initiation

$$(C_6H_5CH_2)_3ZrCH_2C_6H_5 + CH_2{=}CHC_6H_5 \rightleftharpoons \overset{\overset{\displaystyle CH_2{\downarrow}CHC_6H_5}{}}{(C_6H_5CH_2)_3ZrCH_2C_6H_5} \quad [19]$$

(XVI)

$$\overset{\overset{\displaystyle CH_2{\downarrow}CHC_6H_5}{}}{(C_6H_5CH_2)_3ZrCH_2C_6H_5} \xrightarrow{\text{slow}} (C_6H_5CH_2)_3Zr(\underset{\displaystyle C_6H_5}{CHCH_2})CH_2C_6H_5 \quad [20]$$

Propagation

$$(C_6H_5CH_2)_3Zr(\underset{\displaystyle C_6H_5}{CH{-}CH_2})_nCH_2C_6H_5 + CH_2{=}CHC_6H_5$$

(XXIII)

$$\rightleftharpoons \overset{\overset{\displaystyle CH_2{\downarrow}CHC_6H_5}{}}{(C_6H_5CH_2)_3Zr(\underset{\displaystyle C_6H_5}{CH{-}CH_2})_nCH_2C_6H_5} \quad [21]$$

(XXIV)

$$\overset{\overset{\displaystyle CH_2{\downarrow}CHC_6H_5}{}}{(C_6H_5CH_2)_3Zr(\underset{\displaystyle C_6H_5}{CH{-}CH_2})_nCH_2C_6H_5} \xrightarrow{\text{fast}} (C_6H_5CH_2)_3Zr(\underset{\displaystyle C_6H_5}{CH{-}CH_2})_{(n+1)}{-}$$

$$CH_2C_6H_5 \quad [22]$$

The difference in velocities between process 21 and process 22 probably derives from the difference in structure of (XVI) and (XXIV), in the former the $Zr-C-C_6H_5$ is distorted due to interaction of the phenyl group with metal from the normal tetrahedral angle of 109° to 90° (19) as shown in (XXV). Interaction between the adjacent phenyl group on the polymer chain and metal atom in (XXII) is probably prevented because of the polymer chain is attached to the α-carbon atom.

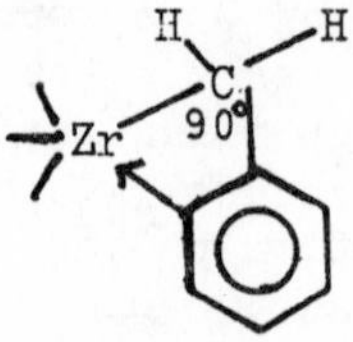

(XXV)

Insertion of the monomer in the $Zr-CH_2C_6H_5$ bond in (**XXV**) will therefore require additional energy, equal to the interaction energy of the phenyl group with the metal atom, since transition metal benzyl compounds are stabilized by the interaction of the aromatic nucleus with the metal atom.

The difference in activity between metal carbon bonds in (XXIV) and (XVI) could also explain why only one benzyl group per metal atom is used. On kinetic grounds alone, quite apart from other objections to two polymer chains growing from one metal atom, the probability of a second benzyl group being displaced can only be comparable to the probability of chain termination.

The above processes are conveniently written in the following form for the purposes of calculation.

$$
\begin{array}{lll}
C + M \rightleftharpoons E & K_1 & (a) \\
E \longrightarrow D_1 & k_1 & (b) \\
D_n + M \rightleftharpoons E_n & K_2 & (c) \\
E_n \longrightarrow D_{n+1} & k_2 & (d) \\
D_n \longrightarrow P_n + F & k_3 & (e) \\
D_n + M \longrightarrow Z_n & k_4 & (f)
\end{array}
\qquad [23]
$$

In these equations E, E_n ($n \geqslant 1$), D_n ($n \geqslant 1$) and F are (XVI) (XXIV), (XXIII) and $(C_6H_5CH_2)_3ZrH$ respectively. C, M, P_n and Z_n represent the catalyst, monomer, polymer and polymer molecules attached to inactive metal atoms respectively.

We drive immediately from these equations the following relationship:

$$
[C]_0 = [C] + \sum_1^\infty [D_n] + \sum_1^\infty [Z_n] + [F] + \sum_1^\infty [E_n] \qquad [24]
$$

It has been shown experimentally that $\sum_1^\infty [D_n] + \sum_1^\infty [E_n] + \sum_1^\infty [Z_n]$ (equal to $Q \approx 10^{-5} M$) and [F] ($\approx 10^{-4} M$) are small compared to the initial concentration of $Zr(benzyl)_4$. In all equations therefore we can put $[C]_0 = [C]$ at low conversions.

Since the equilibrium studies have shown that $[E] << [C]_0$, it is reasonable to assume that $\sum_1^\infty [E_n] << \sum_1^\infty [D_n]$ we therefore write from equations (23a) and (23c):

$$
[E] = K_1[C][M] = K_1[C]_0[M]_0 \qquad [25]
$$

$$
\sum_1^\infty [E_n] = K_2 \sum_1^\infty [D_n][M] \qquad [26]
$$

The rate of initiation, propagation and termination are given by:

$$R_i = k_1[E] = k_1 K_1[C][M] \qquad [27]$$

$$R_p = k_2 \sum_1^\infty [E_n] = k_2 K_2 \sum_1^\infty [D_n][M] \qquad [28]$$

$$R_t = (k_3 + k_4[M]) \sum_1^\infty [D_n] \qquad [29]$$

The rate of polymerization R is equal to R_p because the amount of monomer consumed by the initiation and termination processes compared to the propagation reaction is a trivial quantity. The stationary state principles can be applied to the quantity $\sum_1^\infty [D_n]$, since its concentration is small compared to $[C]_o$ and $[M]_o$ and constant with time at the conversions studied:

$$\sum_1^\infty [D_n] = k_1 K_1 [C]_o [M]_o/(k_3 + k_4[M]_o) \qquad [30]$$

The polymerization rate is therefore from [28] and [30]:

$$R = (K_1 K_2 k_1 k_2 [C]_o [M]_o^2)/(k_3 + k_4[M]_o) \qquad [31]$$

and this agrees with the experimental eq. 13.

The number average degree of polymerization of the isolated polymer $\overline{P}_n$ obtained from [28] and [29] is:

$$\overline{P}_n = R_p/R_t = k_2 K_2 [M]_o/(k_3 + k_4[M]_o) \qquad [32]$$

which is identical with the experimental eq. 14. Comparing eqs. 31 and 32, it is evident that ratios of the intercept and slope in Fig. 3 and 6 should have the same value and be equal to the ratio k_4/k_3. At 30°C the kinetic data gives a value of k_4/k_3 of 0.32 and molecular weight data a value of 0.29. This is very good agreement considering the experimental difficulties and the differing nature of the techniques employed. Finally, division of eq. 31 by eq. 32 gives:

$$R/P_n = k_1 K_1 [C]_o [M]_o = R_i \qquad [33]$$

Polymerization of styrene with a fourfold change in the concentration of monomer   is shown in Table 4. Eq. 28 therefore describes reasonably the change in rate with monomer concentration.

Measurements of $\sum_1^\infty [D_n]$ have been caried out in the case of styrene and ethylene systems and the results are summarized in Tables 5 and 6. A notable feature is that this quantity is small and that only one atom of Zr in about 10^2-10^3 is used at any one time in polymer growth, which is comparable in order of magnitude with the homogeneous model. The closeness of the styrene and ethylene values for $\sum_1^\infty [D_n]$, particularly

TABLE 4.

Relevant constants for the polymerization of styrene initiated by Zr(benzyl)$_4$ in toluene(13).[a]

	30°C	40°C	ΔE,[b] Kcal.
$k_1 K_1$, $M^{-1} sec^{-1}$	3×10^{-8}	6.7×10^{-8}	15
$k_2 K_2$, $M^{-1} sec^{-1}$	0.07	0.2	20
K_1, M^{-1}	~ 0.02	---	--
k_1, sec^{-1}	1.5×10^{-6}	---	--
k_3, sec^{-1}	1.7×10^{-4}	4.0×10^{-4}	16
k_4, $M^{-1} sec^{-1}$	0.44×10^{-4}	2.0×10^{-4}	25
$\sum_1^\infty [D_n]$, M	0.7×10^{-5}	0.6×10^{-5}	

[a][styrene]$_o$ = 3 M, [C]$_o$ = 0.03 M. [b]ΔE (overall) = 16.6 Kcal.

since the latter polymerizes 10,000 times faster, suggests that the measurements are independent of rates and type of monomer. Values of $\sum_1^\infty [D_n]$ in Table 5 show that the number of propagating centers remains more or less constant with conversion suggesting that all chains attached to metal atoms are actively growing. This was not the case with the homogeneous model.

In the absence of transfer agent molecular weights of the polymers formed tend to be very high, chain termination does occur, however, principally by β-hydrogen abstraction. Molecular weights are independent of conversion as shown by the data for polystyrene made with these catalysts and summarized in Table 6. It is also evident from Table 7 that the molecular weight is markedly dependent on the monomer concentration. This was also observed in the homogeneous polymerization of styrene by Zr(benzyl)$_4$, from which $K_1 k_1$ can be obtained. The data from Figs. 3 and 6 for 30°C are combined to give values of $R/\overline{P}_n$ and plotted against [M]$_o$ in Fig. 8. A good straight line is obtained as predicted by eq. 33; from the slope we obtain the value of $K_1 k_1 = 3 \times 10^{-8}$ M^{-1} sec^{-1}

TABLE 5.

Polymerization of styrene and ethylene using $Zr(benzyl)_4/Al_2O_3$ (8).

$[M]_0$ M	$[C]_0$ M	$10^5 \cdot \sum_1^\infty [D_n]$ M	$\sum_1^\infty \dfrac{[C]_0}{[D_n]}$	Temp. °C	$10^5 \cdot R$ $M \cdot sec^{-1}$	$10^5 \cdot R/[M]_0$ sec^{-1}	$k_2 K_2$ $M^{-1} sec^{-1}$
Styrene							
0.5	0.01	---	---	30	0.56	1.12	---
1.0	0.01	0.4	2500	30	0.9	0.9	2.2
2.0	0.01	0.6	1670	30	2.5	1.25	2.1
Ethylene							
0.055	(4.5×10^{-5})	0.0225	200	80	27.0	492	21,800

244

TABLE 6

Polymerization of styrene at 30°C in toluene $[C]_o = 10^{-2}$ M (8).

$[M]_o$ M	Conversion %	10^{-6} M_n	10^5 R_p M·sec^{-1}	10^{-5} $\sum\limits_1^\infty [D_n]$ M
2.0	2.7	1.8	2.5	0.30
	4.3	1.8	2.5	0.60
	7.4	1.6	2.5	0.70
	8.2	2.0	2.5	0.60
1.0	4.1	0.8	0.9	0.3
	8.8	0.7	0.9	0.4
	11.3	0.6	0.9	0.6

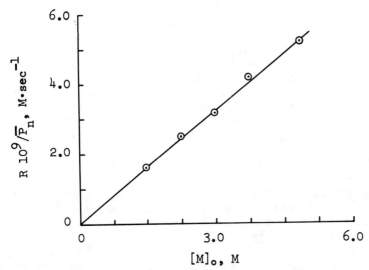

Fig. 8. Polymerization of styrene by Zr(benzyl)$_4$. Data derived from Figs. 3 and 7 plotted in accordance with eq. 28 (6).

at 30°C. The half-life of the catalyst may be shown to be $\ln 2/k_1 K_1 [M]_0$ and is equal to 2000 hours at 30°C when $[M]_0 = 3.0$ M. The catalyst therefore reacts extremely slowly and if there was sufficient monomer available it would require over a year to react completely.

Eq. 23 enables the product $k_2 K_2$ to be obtained using the value of $\sum_1^\infty [D_n]$, the intercept in Fig. 7, and rate data from Fig. 3; at 30°C it is equal to 0.07 $M^{-1}sec^{-1}$.

With this information it is now possible to derive values of k_3 and k_4 from the appropriate slope and intercept in Fig. 3 with the aid of eq. 31. All these constants are listed in Table 4, together with the other constants derived above. In Table 4 data is also given for the polymerization process at 40°C but is probably less accurate since fewer measurements of P_n and Q were obtained at this temperature. The values for the energy of activation obtained for the termination constants therefore can only be considered approximate. It is evident from these, however, that because k_4 increases more rapidly with temperature than $k_1 K_1$ or $k_2 K_2$ that both $\sum_1^\infty [D_n]$ and $\overline{P}_n$ will decrease with increasing temperature. This is not clear from the experimental data because the temperature difference is too small although the trend is discernible. It also follows from eq. 28 that since R is the product of two terms, k_2, which increases with temperature, and $\sum_1^\infty [E_n]$, which decreases with increasing temperature, the rate versus temperature curve will have a maximum.

The values of the equilibrium constant given in Table 2 are probably similar to the benzyl derivative, an approximate value for k_1 can be obtained therefore and is $1.5 \times 10^{-6} sec^{-1}$. Complex (V) has therefore a half-life of approximately 130 hours, confirming that the insertion of the olefin between the $Zr-CH_2\phi$ bond is extremely difficult. Direct measurement of K_2 is not possible. It is a reasonable assumption however that K_2 is within an order of magnitude of K_1 because the environment of the metal in (XVI) and (XXIV) are similar. If we put $K_1 = K_2$ we derive $k_2 \simeq 3.5$ sec^{-1}, the half-life of species (XII) is therefore $\simeq 0.2$ sec.

It is also probable that the polymerization of methyl methacrylate initiated by chromium allyls(21,20) proceeds by a mechanism similar to that described for the polymerization of styrene by $Zr(benzyl)_4$ (2).

5) *KINETICS OF THE POLYMERIZATION OF STYRENE AND ETHYLENE USING HIGHLY ACTIVE CATALYST OF LONG LIFE (8).*

Kinetic studies of the polymerization of ethylene and styrene with heterogeneous catalysts is simplified by using

the homogeneous polymerization of styrene by Zr(benzyl)$_4$ as a
model. In the latter system, eq. 23 relates the rate of polymerization to two measurable parameters $\overset{\infty}{\Sigma}[D_n]$, the number of
propagating centers, and $[M]_o$ the monomer concentration. In
the heterogeneous system D_n corresponds (XXVI) ($n \geqslant 0$) and is
not related in any known way to the number of Zr-atoms
initially added. Values of D_n in heterogenous polymerization
can be measured using the tritium tagging technique already
described for Zr(benzyl)$_4$/styrene.

In Fig. 8 are shown typical changes in polymerization
rate, R, with time obtained for the polymerization of ethylene
using Zr(benzyl)$_4$/Al$_2$O$_3$ catalysts. These steady rates can be
maintained over long periods provided mass transfer problems
do not interfere. The dependence of rate on monomer concentration is given in Fig. 9 and is seen to be, over the range
of concentrations studied, accurately first order.

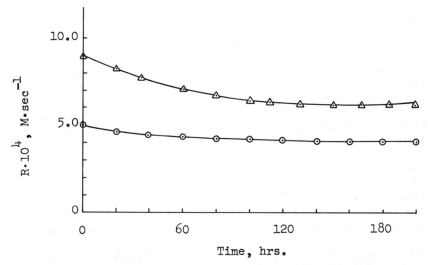

*Fig. 9. Polymerization of ethylene by Zr(benzyl)$_4$/Al$_2$O$_3$ in
toluene(20). Initial catalyst concentration $[C]_o = 1.5$
• 10^{-4} Zr atoms per litre. Pressure 4 atmospheres, equal
partial pressures of ethylene and hydrogen. ($\triangle$) 40°C,
(O) 25°C.*

Direct evidence of the insertion between the zirconium-
metal bond was obtained from work with catalysts prepared
from Zr(C^{14}H$_2$C$_6$H$_5$)$_4$ and is summarized in Table 7. It will be
seen that the amount of polymer derived from the initial
insertion reaction remains stationary but the proportion of
the total declines as the amount of polymer originating from

247

TABLE 7

Polymerization of styrene by $Zr[C^{14}H_2C_6H_5]_4/Al_2O_3$ in toluene at $30°C$; $[M]_o = 1.0\ M$; $[C]_o = 1.0\ x\ 10^{-2}M$ (8).

Conversion %	$10^6\ M_n$	Concentration of C^{14} Polymer $\cdot 10^5\ M$	Fraction of Total of C^{14} -Polymer chains, %
6.8	0.53	0.35	34
14.7	0.64	0.39	18

the re-alkylated Zr-H centers becomes more important. The latter centers are responsible for the conversion of the majority of monomer to polymer in the ethylene system and the initial stage is even less important.

It is not possible to measure the very high molecular weights of polyethylene prepared with these catalysts in the absence of hydrogen. In the presence of hydrogen the values of $\overline{M}_n$ obtained are sensitive to both the concentration of ethylene and hydrogen. The data obtained is summarized in Table 8 and plotted in Figs. 10 and 11. This gives a very good straight line over a wide range of hydrogen and ethylene concentrations.

TABLE 8

Molecular weights of polyethylenes prepared in the presence of hydrogen at $80°C$ in toluene, $[C]_o = 4.5 \cdot 10^{-5}\ M^{-1}$ (8).

$10^3 \cdot [H_2]$ M	$10^2 \cdot [C_2H_4]$ M	$\overline{M}_n$
2.25	2.1	14,200
5.03	3.8	13,400
9.00	5.0	12,400
12.50	6.0	11,400
16.25	7.0	10,600
20.00	7.7	9,000
23.25	8.4	8,000

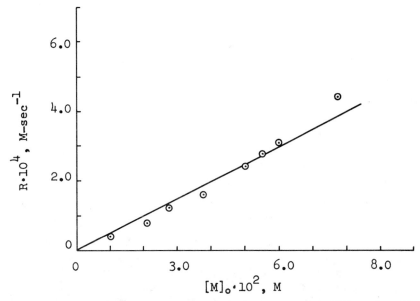

Fig. 10. Dependence of rate on monomer concentration for the polymerization of ethylene at 80°C in toluene $[C]_o = 4.5 \cdot 10^{-4}$ *Zr atoms per litre (20).*

In order to comprehend the chemistry of the process further it is necessary to trace the behaviour of catalyst particles during the polymerization(8). This is readily deducable from the microscopic studies of the polymer produced. In Fig. 12 it is seen that polyethylene is laid down in cylinders of fairly uniform diameter which wind into helices. In the example quoted the original alumina catalyst has completely disintegrated giving a polymer particle with an increase in diameter of five times. Further analysis with the million volt electron microscope at Harwell shows that these alumina particle fragments are located at the tips of the cylinder as shown in Fig. 12. We can therefore visualize the polymerization process as follows.

The initial alumina particle with attached Zr-allyl groups break up in the early stages of polymerization into particles of 1000 Å or less which form nuclei around which chain growth and crystallization occur. The polymer chains generated fold(23) to give lamellae (polythene single crystals) which stack behind the alumina particle with the long axis of polymer chains parallel to the axis of the cylindrical envelope shown in Fig. 13. Direct evidence that the orientation of polythene chains is along the long axis

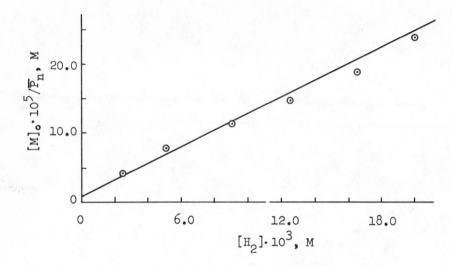

Fig. 11. *Molecular weights of polyethlenes prepared in toluene at 80°C at different hydrogen and ethylene concentrations. Data plotted in accordance with eq. 40.*

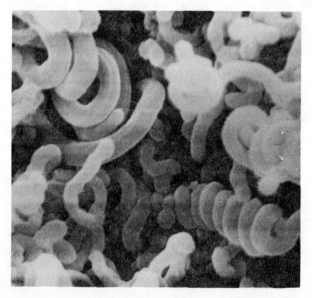

Fig. 12. *Electron micrograph of "as polymerized" polyethylene prepared using conditions described for Fig. 9. Dimensions X10,000 (20).*

of the cylinder in Fig. 12 was obtained by looking at the electron diffraction pattern of a cylinder in corresponding orientation. Well-defined maxima on the (110) and (200) arcs were perpendicular to the cylinder axis, hence the preferred chain direction is seen to be parallel to this axis. Following an initial degradation of the primary catalyst particle, the catalyst fragments are pushed outwards from the center at the tips of cylinders which coil to give helices shown in Fig. 12. It is evident from this description that the diffusion barrier between catalyst sites and monomer is independent of the amount of polymer formed which accounts for the persistence of high rates of polymerization at high conversion (Fig. 9), a characteristic of this system.

The physical constraints leading to the overall morphology however, are not understood. In particular why is such a uniform cylindrical envelope maintained during the growth? One possible explanation is that the polymer at the tip is molten and surface tension effects maintain the shape.

The preceding results suggest that the polymerization process proceeds in accordance with the following mechanism. The catalyst center (XXVI) coordinates monomer giving (XXVII) which is present in low concentrations.

$$
\begin{array}{ccc}
\text{(XXVI)} & & \text{(XXVII)} \\
& + \ CH_2{=}CHR \ \rightleftharpoons & \\
& & \quad [34] \\
& & \text{(XXVII)}
\end{array}
$$

Insertion of the monomer between the benzyl group and metal atom gives (XXVIII). Repetition of this process (eq. 34 and 35) gives high molecular weight polymers. A generalized reaction scheme can be written as follows:

$$
\begin{array}{c}
\text{(XXIX)(n)} \\
\begin{array}{c}
\text{Al-O} \\
| \\
\text{O} \qquad \text{Zr} \\
| \\
\text{Al-O}
\end{array}
\begin{array}{c}
\text{CH}_2\text{C}_6\text{H}_5 \\
\\
\text{(CHR}\cdot\text{CH}_2)_n\text{R}'
\end{array}
\;+\; \text{CH}_2=\text{CHR} \;\rightleftarrows\;
\begin{array}{c}
\text{Al}\cdot\text{O} \\
\text{O} \\
\text{Al}\cdot\text{O}
\end{array}
\begin{array}{c}
\text{CH}_2\text{C}_6\text{H}_5 \\
| \qquad \text{CH}_2 \\
\text{Zr} \leftarrow \| \\
\qquad \text{CHR} \\
\text{(CHRCH}_2)_n\text{R}'
\end{array}
\text{(XXX)}
\end{array}
$$

[35]

$$
\begin{array}{c}
\text{Al-O} \\
| \\
\text{O} \qquad \text{Zr} \\
| \\
\text{Al-O}
\end{array}
\begin{array}{c}
\text{CH}_2\text{C}_6\text{H}_5 \\
| \\
\\
\text{(CHR}\cdot\text{CH}_2)_{n+1}\text{R}'
\end{array}
$$

(XXXI)

An important method of molecular weight control is the β-hydrogen abstraction process:

$$
\text{(XXXI)} \longrightarrow
\begin{array}{c}
\text{Al-O} \\
| \\
\text{O} \qquad \text{Zr-H} \\
| \\
\text{Al-O}
\end{array}
\begin{array}{c}
\text{CH}_2\text{C}_6\text{H}_5 \\
\\
\\
\end{array}
\;+\;
\begin{array}{c}
\text{R} \\
| \\
\text{HC} = \text{CH(CHR}\cdot\text{CH}_2)_n\text{R}'
\end{array}
$$

[36]

(XXXII)

Realkylation of (XXXII) reactivates the propagating center,

$$
\text{(X)} + \text{CH}=\text{CHR} \longrightarrow
\begin{array}{c}
\text{Al-O} \\
| \\
\text{O} \qquad \text{Zr} \\
| \\
\text{Al-O}
\end{array}
\begin{array}{c}
\text{CH}_2\text{C}_6\text{H}_5 \\
| \\
\\
\text{CH}_2\text{R}\cdot\text{CH}_3
\end{array}
$$

[37]

(XXXIII)

and repetition of reactions 34 and 35 gives a new polymer chain (XXXI) where R' is now $\text{CH}_2\text{R}\cdot\text{CH}_3$.

An important difference between the homogeneous and heterogeneous transition metal alkyl compounds is that the hydride species (XXXII) is an active center and realkylates readily whereas species such as $[\text{C}_6\text{H}_5\text{CH}_2]_3\text{Zr}\cdot\text{H}$ are inactive in solution due to dimerization(13).

In the polymerization of ethylene, hydrogen can be used as a transfer agent giving the hydride (XXXII) and a saturated polymer molecule.

$$\text{(XXXI)} + \text{H}_2 \longrightarrow \text{(XXXII)} + \text{CH}_2\text{RCH}_2(\text{CHR}\cdot\text{CH}_2)\text{R}'$$

[38]

In analyzing the kinetics of the process it is worthwhile noting that the initiation reaction[34] together with realkylation of the hydride [37] consume relatively little monomer compared to the growth reactions [34] and [35]. The analysis is then similar to the homogeneous model but with the termination reactions in accordance with the preceding discussion.

$$
\begin{aligned}
C_A + M &\rightleftharpoons E & K_1 \\
E &\longrightarrow D_1 & k_1 \\
D_n + M &\rightleftharpoons E_n & K_2 \\
E_n &\longrightarrow D_n + 1 & k_2 \\
D_n &\longrightarrow P_n + C_H & k_3 \\
D_n + H_2 &\longrightarrow P_n + C_H & k_4 \\
C_H + M &\longrightarrow C_A & k_5
\end{aligned}
\qquad [39]
$$

C_A is (XXVI) or (XXXIII), $E \ldots E_n$ are species (XXVII) and (XXX) $D_1 - D_n$ represent (XXVII) and (XXXI), P_n is a polymer molecule derived from process 36 or 37; and C_H is (XXXII).

The following relations can be written:

$$[E] = K_1[C_A][M] \qquad [40]$$

$$[E_n] = K_2[D_n][M] \qquad [41]$$

and if K_2 is independent of n, it follows:

$$\sum_1^\infty [E_n] = K_2 \sum_1^\infty [D_n][M] \qquad [42]$$

$$[C_A]_0 = \sum_1^\infty [D_n] + \sum_1^\infty [E_n] + [C_A] + [C_H] \qquad [43]$$

The rate of polymerization gives eq. 28.

$$R = -(d[M]/dt) = k_2 \sum_1^\infty [E_n] = K_2 k_2 \sum_1^\infty [D_n][M]$$

In the homogeneous system $\sum_1^\infty [D_n]$ can be computed from the concentrations of Zr(benzyl)$_4$ and styrene but this is not possible in this case since we cannot calculate the number of propagating Zr-atoms located on the terminal position of the cylinders shown in Fig. 12. It is therefore necessary to use the tritium oxide quenching procedure to measure $\sum_1^\infty [D_n]$ directly. The product $K_2 k_2$, determined from the ratio $R/[M] \sum_1^\infty [D_n]$, will tend to be too small by an amount related to difference in concentration of monomer around the growing polymer tip and that close to the catalyst site within the tip. The values of $K_2 k_2$ quoted in Tables 4 and 5 must therefore be regarded as minimum values.

The number average degree of polymerization is given by the relationship

$$\bar{P}_n = \frac{K_2 k_2 \overset{\infty}{\underset{1}{\Sigma}}[D_n][M]_o}{k_4 ! \overset{\infty}{\underset{1}{\Sigma}}[D_n] \cdot [H_2] + k_3 \overset{\infty}{\underset{1}{\Sigma}}[D_n]} = \frac{K_2 k_2 [M]_o}{k_4 [H_2] + k_3} \qquad [44]$$

In the absence of hydrogen $\bar{P}_n = (K_2 k_2/k_3)[M]_o$, that is there should be a marked dependence of molecular weight on monomer concentration. The limited amount of information in Table 4 supports this statement. It has also been found that hydrogen modifies the molecular weight during the polymerization of ethylene.

Eq. 44 predicts that a plot of $[M]_o/\bar{P}_n$ against $[H_2]$ will be a straight line with slope $k_3/k_2 K_2$ and intercept $k_4/k_2 K_2$. A typical plot is shown in Fig. 11 using the data in Table 8. The intercept is very small and hardly detectable on the scale, confirming that $k_3 \ll k_4$, which is consistent with the observation that in the absence of hydrogen very high molecular polymers are obtained. Allowing for this fact the agreement between eq. 44 and Fig. 11 is very good, and we derive from it $k_4' = 1.25 \times 10^{-2} k_2 K_2$ and $k_3 < 1.5 \times 10^{-5} k_2 K_2$.

6) STEREOREGULAR POLYMERIZATION

Initial studies of the polymerization of propylene with transition metal allyl compounds suggested that this monomer could not be polymerized by any of the soluble catalysts available. Subsequent work(2) has revealed, however, that the propylene polymerization is much more susceptible to impurities, in particular traces of ether which compete with the monomer for the coordination sites. When this and other impurities are removed, weak activity is detected. These results are summarized in Table 9.

With activities at such a low level, however, the possibility exists that the minute amounts of polymer formed are not produced by the pure transition metal compound, but that traces of magnesium present from the synthesis of $Zr(allyl)_4$ acts as a co-catalyst. Particular care was taken therefore to reduce magnesium impurities to less than 10^{-4} mole percent by repeated recrystallization. As in the ethylene case, replacement of allyl groups by halogen atoms produces a significant increase in catalytic activity. $Zr(allyl)Br_3$ again is the most active compound. Also as with ethylene the *in situ* generated compounds obtained by reacting $Zr(allyl)_4$ with trityl chloride are more active. The compounds (X) to (XIII) give the most active polymerization catalysts of the transition metals for propylene polymerization. It follows, therefore, that transition metal alkyl compounds can be obtained which have activities for propylene polymerization

TABLE 9

Polymerization of propylene by transition metal alkyl compounds.[a]

Compound	Structure of site	Activity	Polymers soluble in toluene			Polymers insoluble in toluene		
			Total %	Isotactic %	Syndio-tactic %	Total %	Isotactic %	Syndio-tactic %
$Zr(allyl)_4$		0.0014	36	50	50	64	81	19
$Zr(allyl)_3Cl$		0.03	60	37	63	40	82	18
$Zr(allyl)_3Br$		0.013	44	39	61	56	77	23
$Zr(benzyl)_4$		0.0007	100	68	21	0	--	--
$Ti(benzyl)_4$		0.004	98	75	25	2	--	--
$SiO_2/Zr(allyl)_4$	X	0.16	40	31	69	60	74	26
$Al_2O_3/Ti(benzyl)_4$	X	5.5	30	40	60	70	80	20
$SiO_2/Zr(allyl)_3Br$	XI	2.7	55	45	55	45	84	16
$SiO_2/Zr(benzyl)_3Cl$	XI	1.03	52	47	53	48	71	29
$SiO_2/Zr(benzyl)_4$	X	5.8	30	40	60	70	80	20
$Zr[CH_2 \cdot Si(CH_3)_3]_3]_3/Al_2O_3$	X	6.0	30	40	60	70	80	20

[a] In toluene at 65°C, $p_{C_3H_6}$ = 10 atm. (15,16).

comparable with Ziegler catalysts. It has also been shown
that higher olefins such as 4-methyl-pentene-1 styrene, etc.,
readily polymerized by compounds such as (X) to (XIII).

A mixture of soluble and insoluble polypropylenes was
obtained with these catalysts. The insoluble materials were
crystalline solids, with melting points in the region 155-
163°C. Crystallinities measured by X-ray and DTA were in the
region of 40%. The proportion of isotactic and syndiotactic
placements can also be determined using the 220 MHz proton
NMR spectrometer(25) and typical spectra are reproduced in
Fig. 13. It is therefore possible to obtain a quantitative
estimate of the percentage of isotactic and syndiotactic

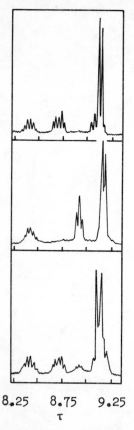

8.25   8.75   9.25
τ

*Fig. 13.   220 MHz proton NMR spectra
of solution of polypropylene (a)
isotactic(25), (b) Syndrotactic
(25), (c) soluble polypropylene
obtained by polymerization with
Zr(benzyl)₄ in solution at 0°C
in a sealed tube.*

placements in both the soluble and insoluble polymers. It is
probably helpful to point out that soluble, non-crystalline
isotactic polypropylene is possible if the runs of isotactic
placements are less than 100 Å in length; blocks of this
length cannot crystallize because molecular motions shake

them apart. The analytical data for polymers obtained with these catalysts is summarized in Table 2. These results indicate that soluble $Zr(allyl)_4$ and $Zr(allyl)_3Cl$ can produce soluble isotactic polypropylene. It is reasonable to assume, therefore, that since the catalysts are also soluble that this is a homogeneous process, that is, a single transition metal atom is able to form isotactic polymers and does not require the environment of a solid surface.

It has been suggested however that isotacticity derives from polymerization occurring on colloidal particles formed by thermal decomposition of the catalysts. As stated previously, in the presence of the monomer even the allyl compounds are stable at 65°C and none of the thermal decomposition products (black to yellow solids) could be detected. As a check on these results a polymerization of propylene was carried out with $Zr(benzyl)_4$ in toluene at 0°C in a sealed tube. The reaction was very slow and analytical quantities of polymer could only be obtained after 312 hours. NMR analysis showed peaks assignable to isotactic sequences, and these were much stronger than the peaks assignable to syndiotactic diads. It was concluded therefore that the polymer obtained was <u>predominantly isotactic.</u> Giannini, Zucchini and Albizzati(26) have also polymerized propylene at low temperatures (20°C) using transition metal benzyl compounds and obtained crystalline polymer with soluble catalysts.

The similarity in product composition obtained with $Zr(allyl)_3Br$ and $Zr(allyl)_3Br/SiO_2$, despite very large differences in activity, suggest that the environmental changes of the metal atom leading to increased rates of polymerization, does not affect the process controlling the microtacticity of the polymer produced. In compounds (X) to (XIII) each zirconium atom is 10 Å from its nearest neighbour. This is too great a distance for there to be cooperation between propagating centers in controlling the stereoregularity of the insertion process. This confirms that individual transition metal atoms can generate isotactic placements.

Equally reactive catalysts for propylene can be prepared from $Zr[CH_2Si(CH_3)_3]_4$ as shown in Table 9. The observations reported here therefore would appear to be quite general.

The polymerization of styrene in toluene, on the other hand, gives exclusively soluble polymer. Crystalline polystyrene is obtained using $Zr(benzyl)_4/Al_2O_3$ catalysts.

7) A GENERAL MECHANISM

Recently an analysis of the common features of homogeneous processes involving an organometallic reaction site
has been carried out(1). It has been adduced that the chemistry is dominated by two processes, namely, β-hydrogen abstraction and olefin insertion. Also, since essentially the same
fundamental reactions occur with Al^{III}, Ti^{IV}, Zr^{IV}, Hf^{IV} (no
d-electrons) as well as with metals such as Pd, Pt, Co (rich
in d-electrons), it is concluded that the nature of the metal
is not critical provided it is sufficiently electron deficient
to form an olefin complex by overlap of s-orbitals of the
metal with the π-orbitals of the olefin. Thus in the case of
Zr and Hf tetrabenzyl, it has been shown that the amount of
complex present in the polymerization process is relatively
small, but non-the-less very high molecular weight polymers
($>10^6$) are formed.

In this context it is worth comparing aluminum alkyl
chemistry and that described previously for transition metal
alkyls. Aluminum alkyls form complexes with amines, ethers
and olefins. Internal olefin complexes have been identified
spectroscopically (27), i.e. (XXXIV) shows a pronounced shift

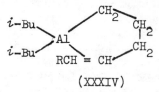

(XXXIV)

(by 30-38 cm^{-1}) to higher wave numbers compared to the corresponding diolefin $CH_3CH=CH \cdot [CH_2]_4 CH=CH_2$ from which the metal
alkyl was formed. The etherates of these aluminum alkyls
show no such frequency shifts as expected. We may conclude
from this that olefin alumina alkyl complexes are formed, and
like the transition metal alkyl compounds the concentration
of the intermolecular complexes is small. It can also be
demonstrated that olefin insertion and β-hydrogen abstraction
are important reaction in aluminum alkyl chemistry and is
frequently part of an equilibrium process. This $(i-Bu)_3Al$
readily forms isobutene and a metal hydride at 120°-150°C,
which recombine at 60°-70°C. The overall processes may be
written(24)

$$(i-Bu)_3Al \;\rightleftarrows\; (iBu)_2 \, AlH + CH_2=C(CH_3)_2 \qquad [45]$$

A similar argument has been applied to the isomerization
of alkyl boranes(28). Analogies of this type can be
usually criticized on the grounds that catalysts which yield

dimers and oligimers do not give high molecular weight polymers. The fact is that there is little difference between
these two types of catalyst so far as the chemistry of the
processes are concerned, they need only differ in velocity
with which olefin insertion relative to β-hydrogen abstraction
occurs; this behaviour is summarized by eq. 32 and 44 which
state that the degree of polymerization is the ratio of the
two competing reactions. Thus we have seen that progressively
replacing three of the allyl groups in $Zr(allyl)_4$ with bromine
atoms gives the oligomerization catalyst $Zr(allyl)Br_3$ whereas
the original compound, in the absence of hydrogen, gives polyethylene of almost infinite molecular weight. Similarly, it
has been found(29) that $(i-Bu)_3Al$ reacts with the hydroxyl
groups of alumina to give a compound analogeous to (XIII) in
which Al replaces Cr. The latter is a reasonably active
catalyst for the polymerization of ethylene and polymers with
molecular weight in excess of 10^6 are obtained. These experiments show that provided the appropriate ligands are chosen
there is no constraint on a non-transition metal behaving as
co-ordination anionic-polymerization catalyst.

Since olefin insertion between the metal hydride bond
and β-hydrogen abstraction are part of an equilibrium process,
they will have a common transition state. The latter process
gives important information on the structure of this intermediate since the products formed suggest only one reaction
pathway, namely:

$$(L)_n Me\!\!\begin{array}{c} H\cdot CHX \\ \diagup \\ CHR \end{array} \rightleftharpoons (L)_n Me\!\!\begin{array}{c} \cdots H \cdots \\ \diagup\quad\diagdown \\ \cdots CHR \cdots \end{array}\!\!CHX \rightleftharpoons (L)_n Me\!\!\overset{H}{\underset{CHR}{+\!\!\mid\!\!\mid}}CHX \rightarrow (L)_n Me\!\!\overset{H}{\underset{CHR}{+\mid\mid}}CHX \quad [46]$$

$$(XXXVa)$$

The olefin insertion reaction is the reverse of process [46]
and we generalize it to include metal alkyls $(L)_n MeX$ as well
as metal hydrides $(L)_n Me\cdot H$, as shown in eq. 47:

$$\begin{array}{c} CH_2 \\ \mid\mid \\ CHR \end{array} + (L)_n Me\!\!\overset{X}{\underset{}{\mid}} \rightleftharpoons (L)_n Me\!\!\overset{X}{\underset{CHR}{+\!\!\mid\!\!\mid}}\!\!CH_2 \rightleftharpoons (L)_n Me\!\!\begin{array}{c} \cdots X \cdots \\ \diagup\quad\diagdown \\ \cdots CHR \cdots \end{array}\!\!CH_2 \rightarrow (L)_n Me\!\!\begin{array}{c} X \diagdown \\ \diagup\quad\diagdown \\ CHR \diagup \end{array}\!\!CH_2$$

$$(XXXVb) \hspace{4cm} [47]$$

In eq. 46 and 47, R = H, alkyl, aryl; X = H, $-(CHR\cdot CH_2)_n R'$
n > 0; Me is a transition metal, Ti, Zr, Cr, etc., or an
electron deficient metal such as Al. L can be any of the
following ligands, $CH_2Si(CH_3)_3$, $CH_2\phi$, allyl, $Al-O^-$, $Si-O^-$,
halogen, etc. The intermediate (XXXV) is probably structur-
ally similar to the π-allyl compounds(3). An important feature to this structure is that three carbon atoms lie in a

plane which is almost at right angles to the plane of the
metal atom (XXXVI).

$$H_2C\text{———}CH_2$$
$$H_2C\cdots M$$

(XXXVI)

and is a resonating structure such that each C-C bond is
intermediate in length between C=C and C-C bond. It is a
characteristic of all organometallic systems of the π-bonded
type that the metal is not in the same plane as the organic
ligands, and resonance occurs between carbon atoms, other
examples are ferrocene and chromocene. The particular importance of the structure of XXXVI is that assymetric synthesis
is more readily comprehended in terms of the geometry of the
center then is the case in which the metal and carbon atoms
are coplanar. The limitations of the coplanar 4-center system
proposed initially by Brown(30) and Egger(31) has been discussed by Streitweiser, Verbit, and Bitman taking as example
the formation of optically active 1-Butanol-1-d from the
optically active borohydride(34). Without further explanation
we propose that the structure of the propagating center and
therefore one of two types XXXVII for isotactic polymerization,
and XXXVIII for syndiotactic polymerization.

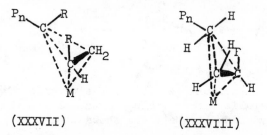

(XXXVII)   (XXXVIII)

Details of this mechanism will be discussed elsewhere.

Finally, it is interesting to compare the data obtained
for these catalysts with radical and anionic propagation processes. Comparable data is summarized in Table 10. It is
evident that propagation rates for these catalysts are at
the high end and similar to the polymerization of free
anions in solution.

TABLE 10.

Bimolecular rate constants for polymerization by various species.

	Temp.	k_p or k_2K_2 $M^{-1}sec^{-1}$
Styrene		
Radical	25°C	27 (32)
Anionic	25°C	∿55,000 (33)
Zr(benzyl)$_4$/Al$_2$O$_3$	25°C	2.0 (8)
Ethylene		
Zr(benzyl)$_4$/Al$_2$O$_3$	80°C	21,800 (8)

LITERATURE CITED

(1) D.G.H. Ballard, *Chemistry in Britain* 10 No. 1, 20 (1974).
(2) D.G.H. Ballard, *Advances in Catalysis* 23, 263 (1973).
    D.G.H. Ballard, 23rd Int. Congr. Pure Applied Chem.,
    Spec. Lect. 6, 219 (1971).
(3) H. Deetrich, and R. Uttech, *Naturwiss.* 50, 613 (1963);
    Z. *Krist* 122, 60 (1965).
(4) D.G.H. Ballard, E. Jones, T. Medinger, and A.J.P. Pioli,
    Macromol. Chem. 148, 176 (1971).
(5) B.K. Bower, Dutch Patent No. 7, 201, 366, *J. Am. Chem.
    Soc.* 94, 2512 (1970).
(6) D.G.H. Ballard, N. Heap, B.T. Kilbourn, and R.J. Wyatt,
    Makromol Chemie 170, 1 (1973).
(7) D.G.H. Ballard, and B. Shaw, unpublished.
(8) D.G.H. Ballard, E. Jones, R.J. Wyatt, R.T. Murray, and
    P.A. Robinson, unpublished.
(9) D.G.H. Ballard, E. Jones, A.J.P. Pioli, P.A. Robinson,
    and R.J. Wyatt, British Patent No. 1,314,828.
(10) D.G.H. Ballard, and J.M. Key, unpublished.
(11) R.S. Coffey, J.M. Key, B.T. Kilbourne, and D. Wright,
     U.K. Application No. 26134/72.
(12) A.J.P. Pioli, unpublished results (1971); reported by
     D.G.H. Ballard in *Advances in Catalysis* 23, 263 (1973).

(13) J.M. Key, unpublished results.

(14) D.G.H. Ballard, D.R. Burnham, and D. Shepherd, unpublished.

(15) R.J. Cventanovic, F.J. Duncan, W.E. Falconer, and R.S. Irwin, *J. Amer. Chem. Soc.* *87*, 1827 (1965).

(16) V. Schung, E. Gilav, *Chem. Commun.* 650 (1971).

(17) L. Vaska, *Inorg. Chem. Acta* *5*, 295 (1971).

(18) R.S. Nyholm, and M. Aresta, *Chem. Commun.* 1459 (1971).

(19) G.R. Davies, J.A. Jarvis, B.T. Kilbourn, and A.J.P. Pioli, *Chem. Commun.* *677*, 1511 (1971).

(20) J.J. Felten, and W.P. Anderson, *J. Organometal. Chem.* *87*, 36 (1972).

(21) D.G.H. Ballard, and T. Medinger, *J. Chem. Soc.* B1177 (1968).

(22) D.G.H. Ballard, and J.V. Dawkins, *Macromol. Chem.* *148*, 195 (1971).

(23) D.T. Grubb, A. Keller, and G.W. Groves, *J. Mat. Sci.* *7*, 131 (1972), *ibid.* *7*, 822, (1972).

(24) G.E. Coates, M.L.H. Green, K. Wade, "Organometallic Compounds", Vol. 1, Methuen & Co. Ltd., London, 1967.

(25) F.A. Bovey, "Polymer Conformation and Configuration", Academic Press, New York, p. 36 (1969).

(26) U. Giannini, U. Zucchini, and E. Albizzati, *Polymer Letters* *8*, 405 (1940).

(27) G. Hata, *Chem. Commun.* 7 (1968).

(28) F.M. Rossi, P.A. McCusker, and G.F. Hennion, *J. Org. Chem.* *32*, 450 (1967).

(29) D.G.H. Ballard, and D. Burnham, unpublished.

(30) H.C. Brown, and G. Zwerfel, *J. Amer. Chem. Soc.* *83*, 2544 (1961).

(31) K.W. Egger, *J. Amer. Chem. Soc.* *91*, 2867 (1969).

(32) M.S. Matheson, E.E. Auer, E.B. Bevilacqua, and E.J. Hart, *J. Amer. Chem. Soc.* *73*, 1700 (1951).

(33) D.J. Worsfold, and S. Bywater, *J. Phys. Chem.* *70*, 162 (1966).

(34) A. Streitwieser, L. Verbit, and R. Bittman, *J. Org. Chem.* *32*, 1530 (1971).

A Kinetic Approach to Elucidate the Mechanism of Ziegler-Natta Polymerization

TOMINAGA KEII

Department of Chemical Engineering
Tokyo Institute of Technology
Meguro, Tokyo, Japan

1) INTRODUCTION

To determine the reaction mechanism kinetic studies in heterogeneous reactions are less powerful than those in homogeneous reactions because of the localization of reaction loci on the interface between phases. Sometimes the non-uniformity (heterogeneity) of the surface of solid catalyst makes the kinetic behavior more complex in heterogeneous catalysis. Nevertheless, kinetic studies are valuable for two reasons. The results can be used to guide process design or reactor design. Secondly, kinetic data can also help to eliminate those reaction mechanisms which are incompatible with the kinetics. In the case of Ziegler-Natta polymerization, there are many proposed mechanisms and some of them are widely accepted. On the other hand the kinetic results have been accumulated. In this paper new kinetic results are used to examine some fundamental and practical problems of the Ziegler-Natta polymerization of propylene.

2) PHYSICAL PROCESSES INVOLVED IN SLURRY POLYMERIZATION

Firstly, it should be noted that the olefin polymerizations with $TiCl_3$-organometallic compound catalysts in inert solvent under gas phase, may be affected by some physical processes, even on a laboratory scale. Thus, the overall rate of gas consumption can not be taken as the (kinetic) polymerization rate, unless it is confirmed to be free from any influence of physical process. The most significant physical process involved in these slurry polymerizations is the gas absorption into the slurry. The evaluation of the rate of absorption of gas into the slurry of polymerization system, is difficult in the case of the heterogeneous polymerizations with $TiCl_3-AlR_3$ catalysts, since both the viscosity of the slurry and the catalytic activity change with increasing polymerization time. Developing a new method applicable for these polymerizations, the present author and his co-workers have evaluated the influence of gas absorption in propylene polymerization with $TiCl_3-AlEt_3$ (1,2) and $TiCl_3-AlEt_2Cl$ (3). The method is based on the following theory of the transition phenomena which accompanied sudden changes of the agitation speed during a polymerization. The overall polymerization rate, $R_{t,a}$, at time t and stirring speed a, may be represented by

$$R_{t,a} = \beta_{t,a}([M]_0-[M]_a) = k_t G[M]_a \qquad [1]$$

where $\beta_{t,a}$ is the absorption rate constant at t and a, $[M]_0$ or $[M]_a$ is the monomer concentration in equilibrium with gas phase or that in the slurry, k_t and G the kinetic rate constant of polymerization per unit weight of the catalyst and total weight of the catalyst in the slurry. When the stirring speed of the stirrer changes suddenly from a to b, the absorption rate instantly assumes a new value, $\beta_{t,b}$. However, the monomer concentration changes much more gradually. These transitional phenomena can be expressed as follows,

$$R_{t,a} \xrightarrow{\text{rapid}} R^0_{t,b} \xrightarrow{\text{slow}} R_{t,b} \qquad [2]$$

$$R^0_{t,b} = \beta_{t,b}([M]_0-[M]_a) \qquad [3]$$

$$R_{t,b} = \beta_{t,b}([M]_0-[M]_b) = k_t G[M]_b \qquad [4]$$

From eqs. 1 and 3, we have

$$R^0_{t,b}/R_{t,a} = \beta_{t,b}/\beta_{t,a} \qquad [5]$$

which gives the dependence of β on stirring speed. Based on this relation 5, we have obtained the following expression for the propylene polymerization with $TiCl_3-AlEt_3$-n-heptane in a 500 ml flask reactor with a paddle stirrer

$$Sh \approx 0.25 \cdot Re^{1.0} \cdot Sc^{0.5} \qquad 11 \leqslant Re \leqslant 700 \qquad [6]$$

where Sh is Sherwood's number, Re is Reynold's number and Sc is Schmidt's number, respectively.

Fig. 1 shows the difference between the overall rate of gas consumption (B) and the true polymerization rate (A) which is free from the influence of gas absorption. The true polymerization rate, thus obtained, was independent of the Reynold's number, which strongly shows that it is free from

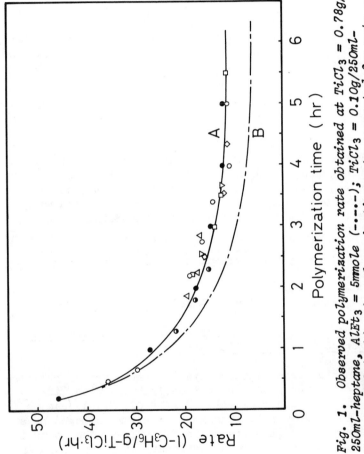

Fig. 1. Observed polymerization rate obtained at $TiCl_3 = 0.78g/250ml$-heptane, $AlEt_3 = 5mmole$ (-·-·-); $TiCl_3 = 0.10g/250ml$-heptane (—●—); kinetic polymerization rate corrected for rate of monomer absorption: $TiCl_3 = 0.28g/250ml$-heptane (▣); $TiCl_3 = 0.34g/250ml$-heptane (●); $TiCl_3 = 0.47g/250ml$-heptane (▽); $TiCl_3 = 0.53g/250ml$-heptane (◇); $TiCl_3 = 0.78g/250ml$-heptane (○); $TiCl_3 = 0.85g/250ml$-heptane (△).

any more influence of other physical process such as the monomer transfer through polymer film on the surface of $TiCl_3$. The relation 5 is useful for polymerizations in which the absorption resistance is so small that it is not explicitly appear. The polymerization activity of $TiCl_3$-$AlEt_2Cl$ catalyst is small compared with that of $TiCl_3$-$AlEt_3$. The rate of gas absorption does not affect the slower polymerizations because

$$B \gg k_t G, \qquad R_t \simeq k_t G[M]_0 \qquad [7]$$

On the other hand, the rate of a fast polymerization at time τ after a sudden change of stirring speed from a to b can be represented by

$$\frac{R_b^\tau - R}{R_b^0 - R} \rightleftharpoons e^{-\frac{\beta_b}{V}\tau} \qquad [8]$$

where V is the slurry volume. From the measured values of R, R_b^0, and R_b^τ, β_b can be calculated. The absorption rate in the polymerization with $TiCl_3$-$AlEt_2Cl$-n-heptane in the 500ml cylindrical glass flask with four baffle plates which was equipped with six-paddle stirrer, has been found to be

$$Sh \simeq const \cdot Re^{0.7} \cdot Sc^{0.5} \qquad [9]$$

The details of the latter experiment will be published elsewhere.

3) KINETIC BEHAVIORS OF HETEROGENEOUS POLYMERIZATIONS

3.1) Kinetic Curves of Heterogeneous Polymerizations with Different Organometallic Compounds

The kinetic curves (polymerization rates versus polymerization time) of the propylene polymerizations with $TiCl_3$-AlR_3 (R = Et, Dec) and $TiCl_3$-$AlEt_2X$ (X = Cl, Br, I) can be classified into two types. One has a maximum followed by a gradual decrease, as illustrated in Fig. 2, and the other is one of monotonic increase to a stationary rate, as shown in Fig. 3 (4). The kinetic curves of the polymerizations with active $TiCl_3$-AlR_3, -$AlEt_2I$ (R_∞ = 0) and -$ZnEt_2$ ($t_m \simeq 0$) belong to the former type, while those obtained with poisoned-$TiCl_3$-$AlEt_3$ catalyst, and active-$TiCl_3$-$AlEt_2Cl$, and -$AlEt_2Br$ catalyst, belong to the latter type. For convenience, we divide a polymerization into three stages: an initial stage during which the rate of polymerization increases, an intermediate stage during which the rate slows down from a maximum to a stationary rate; stationary period.

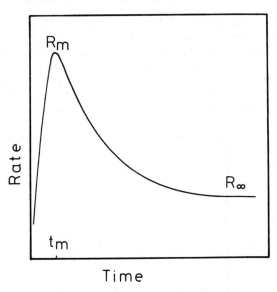

Fig. 2. *Typical kinetic curve*

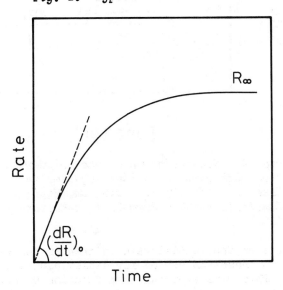

Fig. 3. *Typical kinetic curve*

3.2) Initial Increase of Polymerization Rate

The initial behavior of a propylene polymerization depends upon whether the monomer is added to a premixed catalyst (Case A) or whether the organometallic activator is added to a $TiCl_3$ slurry under propylene (Case B). The initial increase in polymerization rate is gradual for Case A (Fig. 4). There appears to be an induction period when $AlEt_2X$ was used; there was no induction period with AlR_3.

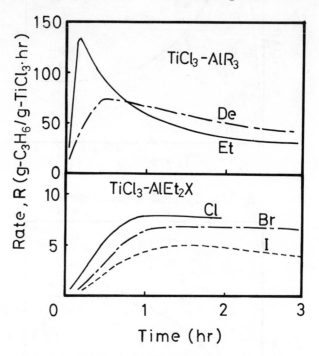

Fig. 4. *Kinetic curve of the polymerization with various catalyst systems in Case A: 41°C, $P_{C_3H_6}$ = 0.87atm, $TiCl_3$ = 5mmole/l, AlR_2Y = 20mmole/l, n-heptane = 250ml.*

The increase of the initial rate is greater for Case B and there was no induction period for dialkylaluminum halides (Fig. 5). This rate increase is a function of both monomer pressure, P, and concentration of the organometallic compound, C. The dependence is not the same for the two cases; the relationships are as follows(5,6)

Case A $\qquad \left(\dfrac{dR}{dt} \right)_0 = k_1 P^2 \left(\dfrac{KC}{1 + KC} \right)$ [10]

Case B $\qquad \left(\dfrac{dR}{dt} \right)_0 = k_2 PC$ [11]

where k, K are constants. The values of the constants and their activation energies are summarized in Table 1.

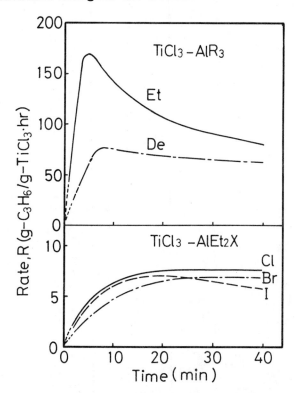

Fig. 5. *Kinetic curves of the polymeriza-*
tion with various catalyst systems in Case
B: experimental conditions are same as for
Fig. 4.

After the initial period, the system attains the same maximum rate of polymerization, R_m, and subsequently the same stationary rate, R_∞ regardless of the order of mixing of reactants. The kinetic expressions for R_m and R_∞ are as follows(6,7)

TABLE 1. *Values of rate constants and their activation energies.*

AlR_2Y	$R_{mA}{}^a$ (g-C_3H_6·g-$TiCl_3{}^{-1}$·hr^{-1}·atm^{-1})	$R_{mB}{}^a$ (g-C_3H_6·g-$TiCl_3{}^{-1}$·hr^{-1}·atm^{-1})	$t_{mA}{}^a$ (min)	$t_{mB}{}^a$ (min)	$R_\infty{}^a$ (g-C_3H_6·g-$TiCl_3{}^{-1}$·hr^{-1}·atm^{-1})	$k_p{}^a$ (g-C_3H_6·g-$TiCl_3{}^{-1}$·hr^{-1}·atm^{-1})	E_p (kcal/mole)	$k_1{}^a$ (g-C_3H_6·g-$TiCl_3{}^{-1}$·hr^{-2}·atm^{-2})	E_1 (kcal/mole)
$AlEt_3$	160	190	9	5	32	35	10	1080	20
$AlDe_3$	75	85	30	8	33	37	11	210	--
$AlEt_2Cl$	--	--	--	--	9.0	12.1	12	--	--
$AlEt_2Br$	--	--	--	--	7.8	9.1	12	--	--
$AlEt_2I$	5.7	8.5	90	16	0	10.4	13	--	--

AlR_2Y	$k_2{}^a$ (g-C_3H_6·g-$TiCl_3{}^{-1}$·hr^{-2}·atm^{-1}·mmole-Al^{-1})	E_2 (kcal/mole)	$k_3{}^a$ (hr^{-1})	E_3 (kcal/mole)	$k_4{}^0{}^a$ (hr^{-1}·atm^{-1})	E_4 (kcal/mole)	K^a (1/mole)	Q (kcal/mole)	I.I. (%)	$\bar{M}_v$ (×10^{-4})
$AlEt_3$	140	17	1.4	2∼3	0.20	8	150	12	80[b]	30.0
$AlDe_3$	40	17	0.4	∼0	0.20	--	160	--	71[b]	--
$AlEt_2Cl$	4.2	17	0	--	1.00	11	110	1	91[c]	50.6
$AlEt_2Br$	2.3	16	0	--	0.62	11	100	0	97[c]	46.6
$AlEt_2I$	3.0	17	0.2	15	0.52	10	130	0	98[c]	30.5

[a] At 41°C, p=66cmHg, $TiCl_3$=16mmol/1, AlR_2Y=20mmol/1. [b] Values determined by infrared method. [c] Fraction of boiling n-heptane insoluble polymer.

$$R_m = k_p'P\left(\frac{KC}{1 + KC}\right) \qquad [12]$$

$$R_\infty = k_pP\left(\frac{KC}{1 + KC}\right) \qquad [13]$$

The gradual decrease in polymerization rate from its maximum to the stationary rate could be expressed by a first order kinetics(8),

$$-\frac{d(R-R_\infty)}{dt} = k\,(R-R_\infty) \qquad [14]$$

The experimental values of the constants k_p, k_p', and k_3 and their activation energies are also given in Table 1. In a latter section the decay of activity of the $TiCl_3$–AlR_3 system will be discussed.

3.3) Kinetic Interpretation; Formation of Polymerization Centers

A probable kinetic interpretation of the initial increase of the polymerization rate is the formation of active centers during the initial stage. Denoting $\overline{k}_p$ as the propagation rate constant per polymerization center, the polymerization rate in the initial stage may be written as

$$R_t = \overline{k}_pP\,[P^*]_t \qquad [15]$$

where $[P^*]_t$ is the number of polymerization centers per unit weight of $TiCl_3$ at time t. Substituting eq. 15 into eqs. 10 and 11, we obtain the rate of formation of P^* for Case A and B to be

Case A $\qquad \left(\dfrac{d[P^*]}{dt}\right)_0 = \dfrac{k_1}{\overline{k}_p}\,P\left(\dfrac{KC}{1 + KC}\right) \qquad [16]$

Case B $\qquad \left(\dfrac{d[P^*]}{dt}\right)_0 = \dfrac{k_2}{\overline{k}_p}\,C \qquad [17]$

Equation 16 for Case A suggests a kinetic model in which the polymerization centers are formed by the reaction of the monomer with organometallic compound dimer adsorbed on the surface of $TiCl_3$. On the other hand eq. 17 for Case B suggests that the polymerization centers are formed by the reaction of organometallic compound dimer with a monomer-covered surface of $TiCl_3$. According to these kinetic models, and making use of the Langmuir isotherm, we derived the theoretical rate of production of active centers. For Case A this is

$$\frac{d[P*]}{dt} = \bar{k}_1[M]S\Theta_A(1-\Theta*)$$

$$\approx \bar{k}_1[M]S\Theta_A \qquad\qquad t \approx 0 \qquad\qquad [18]$$

where S is the total active sites, Θ_A the degree of adsorption of the adsorbed dimers and $\Theta*$ the fraction of the polymerization centers over the adsorbed active sites. For Case B, the rate is given by

$$\frac{d[P*]}{dt} = \bar{k}_2 CS\Theta_M(1 - \Theta*)$$

$$\approx \bar{k}_2 CS\Theta_M \qquad\qquad t \approx 0 \qquad\qquad [19]$$

where Θ_M is the degree of adsorption of monomers. Comparison of eqs. 16 and 18 shows that

$$\Theta_A = \frac{KC}{1 + KC} \qquad\qquad [20]$$

Similarly, eqs. 17 and 19 lead to

$$\Theta_M \approx 1 \qquad\qquad [21]$$

The supposition of the adsorption of organometallic compound in dimer form is based on the form of eq. 20, a Langmuir adsorption type. If only the monomeric form of the organometallic compound is adsorbed, then eq. 20 would instead be

$$\Theta_A = \frac{K\sqrt{C}}{1 + K\sqrt{C}} \qquad\qquad [22]$$

a statement contrary to our experimental results, although Ingberman et al.(9) used this form with a correction of the concentration C. Here, it should be noted that the word "adsorption" has the conventional meaning and significance.

According to these kinetic models, we may arrive at the conclusion that the polymerization centers are formed by the attack of organometallic compound dimer on the sites adsorbed with propylene monomers. In Case A the adsorption of monomers is prevented by the adsorbed organometallic compound and the formation rate of polymerization centers is lower than the rate in Case B. This kinetic model for the formation of polymerization centers has been further supported by the following experiments.

3.4) <u>Rate Decreases caused by the Elimination of the</u>
<u>Organometallic Compound during Polymerization.</u>(5,6)

When the maximum or stationary polymerization rate was
established, the propylene was pumped off and nitrogen was
admitted, the solution filtered out and the residual solid
phase washed several times with pure n-heptane. The same
amount of pure solvent was added, the gas phase was changed
to propylene and then the polymerization is resumed. The
first type of washed catalyst is active even in the absence
of organometallic compound, as illustrated in Fig. 6. On the

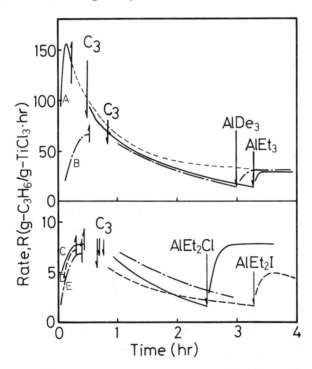

Fig. 6. *Kinetic behaviors of various washed catalysts at*
41°C, $TiCl_3$ = 5mmol/l, $P_{C_3H_6}$ = 0.87atm (A∿E); normal
run (AlR_2Y = 20mmol/l) obtained with $TiCl_3$-$AlEt_3$(A),
-$AlDe_3$(B), -$AlEt_2Cl$(C), -$AlEt_2I$(D), and -$AlEt_2Br$(E),
respectively. The upward arrow indicates removal of
the solution and propylene. The downward arrow with
C_3 indicates addition of pure solvent and propylene,
and one with AlR_2Y addition of organometallic com-
pounds (20 mmol/l).

other hand, a catalyst slurry such as TiCl₃+organometallic compound in *n*-heptane, which was never exposed to propylene and was kept under nitrogen, does not show any polymerization activity after washing (Fig. 7). This second type of washed catalyst can be activated by the addition of organometallic compounds as illustrated in Fig. 7. This experimental result

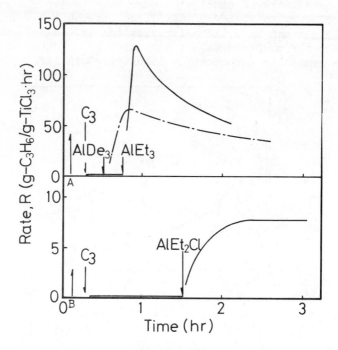

Fig. 7. *Kinetic behaviors of various washed catalyst system at 41°C, TiCl₃ = 5 mmol/l, AlR₂Y = 20 mmol/l. At A or B, solution and propylene were removed after catalyst system (TiCl₃-AlEt₃, -AlDe₃ or -AlEtCl₂) was kept under nitrogen.*

shows that propylene monomer is necessary for the formation of polymerization centers. The reaction of TiCl₃ with organometallic compound without propylene does not produce polymerization centers.

In all the polymerizations with the washed catalysts in the absence of organometallic compound, the polymerization rates decrease, more or less, with increasing of polymerization time (6). This decrease can be expressed by

$$- \frac{dR}{dt} = k_4 R \qquad [23]$$

where $$k_4 = k_4^0 P \qquad\qquad [24]$$

The decrease in the polymerization rate in the cases with the washed catalysts which were obtained from $TiCl_3-AlR_3$ could be expressed by a combination of the two equations, eq. 14 and 23. The values of k_4 depend upon the kind of organometallic compound, as shown in Table 1. This strongly suggests that the polymerization centers formed with different organometallic compounds are not the same. We now present other experimental results which tend to support this conclusion.

3.5) Substitution of Organometallic Compound during Polymerization. (6)

In this experiment, to a washed catalyst of the first type suspended in pure solvent, was added another organometallic compound, II, and the rate of polymerization was followed. The isotacticity of the polymer produced was determined as the fraction (%) of unextractable polymer in boiling n-heptane. The most interesting results have been obtained with $I = AlEt_2I$ or $AlEt_2Br$ and $II = AlEt_2Cl$. The kinetic curves obtained are shown in Fig. 8, which shows that the kinetic curve after the addition of $AlEt_2Cl$ is the same as the usual polymerization with $TiCl_3-AlEt_2Cl$. The

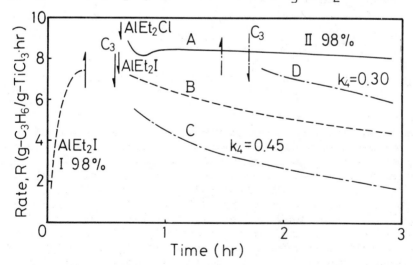

Fig. 8. Substitution of organometallic compound ($I = AlEt_2I$ and $II = AlEt_2Cl$) at 41°C, $TiCl_3 = 5$ mmol/l, $AlEt_2X = 20$ mmol/l. (A) washed catalyst + $AlEt_2Cl$, (B) washed catalyst + $AlEt_2I$, (C) washed catalyst, (D) washed A.

275

kinetic curve obtained in the case with I = AlEt$_2$X and II = AlEt$_3$ show more clearly the change in polymerization rate from that with I to that with II, as illustrated in Fig. 9.

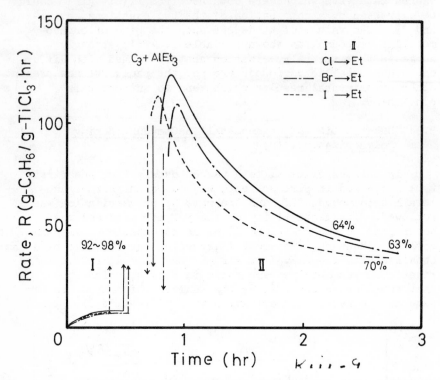

Fig. 9. Substitution of organometallic compound (I = AlEt$_2$X and II = AlEt$_3$) 41°C, TiCl$_3$ = 5 mmol/l, AlEt$_2$Y = 20 mmol/l.

The isotacticity of total polymers isolated after 4∿5 hours of polymerization was 98% for the first case employing in the former I = AlEt$_2$I or AlEt$_2$Br and II = AlEt$_2$Cl. This value is the same as polymers obtained in the conventional experiment with TiCl$_3$-AlEt$_2$I or TiCl$_3$-AlEt$_2$Br. In contrast, TiCl$_3$-AlEt$_2$Cl produces polymers with 92% isotacticity. Since more than 95% of the total polymers was produced after the addition of AlEt$_2$Cl, the higher isotacticity (98%) shows that the stereospecificity of the catalyst is not significantly altered by the addition of II. In the reversed case, i.e. I = AlEt$_2$Cl and II = AlEt$_2$I, the polymerization rate changed into that of the polymerization with II, as before, whereas the isotacticity of the total polymers was 98% which was that of II. The isotacticity of the total polymers in the case with I = AlEt$_2$Cl and II = AlEt$_3$ was 64%. The addition of AlEtCl$_2$ as II causes rapid deactivation of the washed catalyst as illustrated in Fig. 10. The activity can be

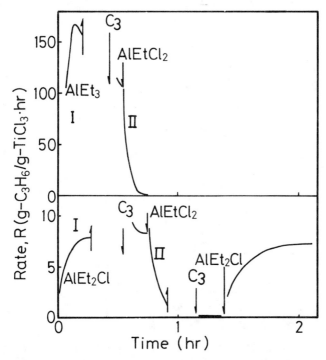

Fig. 10. Substitution of organometallic compound (I = AlEt$_2$Y and II = AlEtCl$_3$) 41°C, TiCl$_3$ = 5mmol/l, AlEt$_2$Y = 20mmol/l.

recovered by the removal of AlEtCl$_2$ and replacing it with AlEt$_2$Cl. This shows that AlEtCl$_2$ is not a cocatalyst. When I = AlEt$_3$ and II = AlEt$_2$Cl, the kinetics is rather complex, as illustrated in Fig. 11. The initial rapid decrease in the kinetic curve is similar to the decrease due to the substitution by AlEtCl$_2$ and is proportional to the amount of AlEt$_2$Cl added. This result shows that the active organometallic compound AlEt$_2$Cl behaves as the inactive one, AlEtCl$_2$, for a portion of the polymerization centers of the washed catalyst obtained from (TiCl$_3$+AlEt$_3$)/n-heptane under propylene.

The experimental results described in this section (Table 2) may be explained on the basis that the polymerization center involves both an organometallic compound and a halide atom in the case of AlEt$_2$X. The former compound controls the polymerization rate and the latter controls the

TABLE 2.

Substitution of organometallic compound during polymerization.

I[a]	II[b]	Isotacticity[c] (%)		Stereospecificity	Kinetic curve
$AlEt_2I$	$AlEt_2Cl$	98 ± 1	(3 runs)	$AlEt_2I$	$AlEt_2Cl$
$AlEt_2I$	$AlEt_3$	70	(1 run)	$AlEt_3$	$AlEt_3$
$AlEt_2Br$	$AlEt_2Cl$	97 ± 1	(2 runs)	$AlEt_2Br$	$AlEt_2Cl$
$AlEt_2Br$	$AlEt_3$	63	(1 run)	$AlEt_3$	$AlEt_3$
$AlEt_2Cl$	$AlEt_2I$	98	(1 run)	$AlEt_2I$	$AlEt_2I$
$AlEt_2Cl$	$AlEt_2Br$	97	(1 run)	$AlEt_2Br$	$AlEt_2Br$
$AlEt_2Cl$	$AlEt_3$	64	(1 run)	$AlEt_3$	$AlEt_3$
$AlEt_2Cl$	$AlEtCl_2$	---		---	$AlEtCl_2$
$AlEt_2Cl$	$ZnEt_2$	---		---	$ZnEt_2$
$AlEt_3$	$AlEt_2Cl$	---		---	$AlEt_2Cl$
$AlEt_3$	$AlEtCl_2$	---		---	$AlEtCl_2$

[a]Initial organometallic compound used on the polymerization for 0.2∿0.5 hr. [b]Organometallic compound added on the washed catalyst (I). [c]Percent insoluble in boiling n-heptane. The polymers given after the addition of II were more than 95 wt. % of total polymers produced.

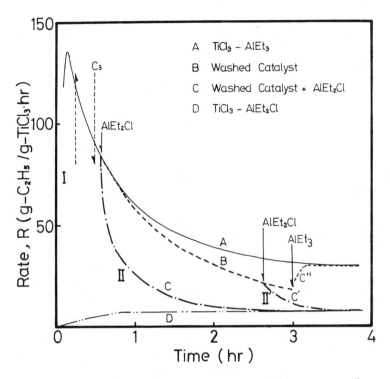

Fig. 11. Substitution of organometallic compound
(I = AlEt₃ and II = AlEt₂Y) 41°C, TiCl₃ = 5mmol/l,
AlEt₂Y = 20mmol/l. (A) usual run with TiCl₃ = 5
mmol/l, AlEt₂Y = 20mmol/l; (B) washed catalyst;
(C;C') washed catalyst + AlEt₂Cl; (D) usual run
with TiCl₃ - AlEt₃.

stereospecificity of the polymerization center. Furthermore,
these species can be easily exchanged with the organometallic
compound dimer in the solution.

3.6) Two-kinds of Polymerization Centers on the TiCl₃ + AlR₃
       Catalyst.

As described in the previous section, the kinetic
curves of the polymerization with TiCl₃ - AlR₃ have gradual
decreasing part from the maximum rate R_m to the stationary
one R_∞. The decreasing rate has been expressed by eq. 14.
The decreasing constant k_3 could be represented in the case
with AlEt₃ as

$$k_3 = k_3^0 exp(-2\ 3kcal/RT) + k_3^{0'} Pexp(-12000/RT) \qquad [25]$$

where the first term is larger than the second term at room temperature and atmospheric pressure. The second term is similar to k_4 in eq. 23, i.e. the decreasing constant of the polymerizations with washed catalysts, in its dependence on the propylene pressure and its value of activation energy. The first term, which is independent of both the concentration of AlEt$_3$ and the propylene pressure, has not been explained. As described before, the decreasing rate of polymerization rate in the polymerization with its washed catalyst in pure solvent can be represented by the combination of eq. 14 and eq. 23. The latter equation corresponds to the decay of the stationary polymerization rate R_∞. After the rate decreased to less than R_∞, the readdition of AlEt$_3$ resulted in the recovery of the polymerization rate to the stationary rate R_∞, as illustrated in Fig. 6.

As reported before(10), the maximum rate of polymerization, R_m, strongly depends on the history of TiCl$_3$ used. Fresh ground samples of TiCl$_3$ give the maximum rate, while unground and old ground samples give lower maximum rates or no maximum at all. The same experimental result has been reported by Natta *et al.* (11). The use of TiCl$_3$ which was aged with a concentrated solution of AlEt$_3$ resulted in the lowering of the maximum rate in accordance with the ageing time(12). Here it should be noted that in all cases the value of R_∞ remained unchanged. This author and his co-workers found that the stationary rate per unit weight of TiCl$_3$ sample fresh or old ground or unground, is proportional to the BET surface area the TiCl$_3$ samples(10,13). These experimental results suggest that there are two kinds of surface sites of TiCl$_3$ as polymerization centers.

Keii *et al.* (14) found that the isotacticity of polymers increases with the decrease of R_m:

$$\overline{I.I.} = I_A + (I_B - I_A) \frac{R_\infty}{\frac{1}{t}\int_0^t R dt} \qquad [26]$$

Fig. 12 shows the reciprocal relation obtained with various catalyst systems. This relation has been confirmed with the experimental results reported by Hoeg and Liebman(15) who carried out the propylene polymerization with TiCl$_3$ – AlEt$_3$ – amine. This reciprocal relation leads phenomenologically only a distribution function of step-type,

$$I.I. = I_A \qquad \text{for } R\text{-}R_\infty$$

$$I.I. = I_B \qquad \text{for } R_\infty \qquad [27]$$

which gives us for the observed isotacticity $(\overline{I.I.})$

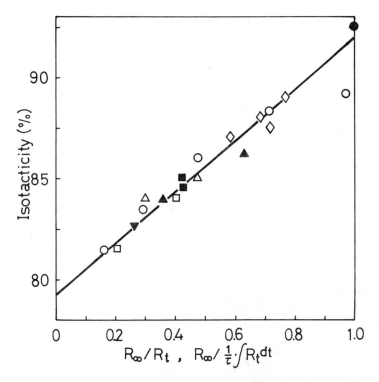

Fig. 12. *Reciprocal relation between isotacticity and reciprocal of polymerization rate (P = 70cmHg at 41°C, and [TAC-α-TiCl₃] = [A] = 16mmol/l in 200ml n-heptane. TiCl₃/AlEt₂Cl, ●; TiCl₃/AlEt₃ (various ageing times),○; TiCl₃ (treated with conc. AlEt₃)/AlEt₃,▲; TiCl₃ (treated with conc. AlEt₂Cl)AlEt₃,▼; TiCl₃ + AlEt₃ (integral),□; TiCl₃ (stored for 2 years)/AlEt₃,△; TiCl₃ (treated with Et₃N)/AlEt₃,■; TiCl₃/AlEt₃, gas-phase polymerization,).*

$$\overline{I.I.} = \frac{I_A \int_0^t (R-R_\infty)dt + I_B \int_0^t R_\infty dt}{\int_0^t Rdt} \qquad [28]$$

$$= I_A + (I_B - I_A)\frac{R_\infty}{\frac{1}{t}\int_0^t Rdt} \qquad [29]$$

Since the part $(R-R_\infty)$ is due to the unstable polymerization centers, as mentioned above, it may be said from eq. 27 that the unstable polymerization centers (denoted by P_A^*) produce low isotactic (I_A) polymers and the stable centers (denoted by P_B^*) which are responsible for the stationary polymerization (R_∞) produce highly isotactic (I_B) polymers. According to the experimental results, Fig. 12, the value of I_B is about 92%, which is the same as that of polymers produced with $TiCl_3$ - $AlEt_2Cl$ catalyst. Therefore, the polymerization centers P_B^* have the same stereospecificity as those of the catalyst $TiCl_3$ - $AlEt_2Cl$. The other kind of polymerization centers, i.e. the unstable ones P_A^*, are postulated as those formed on the surface sites produced by grinding of $TiCl_3$. The surface sites become deactivated upon long standing or ageing with a concentrated solution of $AlEt_3$. Furthermore, the polymerization centers formed on the surface sites deactivated spontaneously with a small activation energy, 2~3 kcal/mole. A key to the structure of these centers is the ESR spectrum, with 1.97, which may indicate a complex of Ti(III) with aluminum alkyl. The disappearance of the spectrum may be due to the conversion of Ti(III) to a complex of lower valence state, such as Ti(II). This hypothesis is in agreement with the experimental observation that the catalyst system $TiCl_3$ - $AlEt_3$ (conc. solution), evolves ethane and has strong activity for H_2-D_2 exchange, but the catalyst system $TiCl_3$ - $AlEt_2Cl$ has neither.(16)

3.7) Polymerization Mechanism from Kinetic Model

The kinetic results given in the preceeding sections can be described by the following model.

(1) There are two kinds of surface sites in $TiCl_3$ which are available as polymerization centers; one is unstable (S_A) and the other stable (S_B). The number of the latter is proportional to the BET surface area. The organometallic compounds AlR_3 convert both to active polymerization centers, whereas $AlEt_2X$ produces polymerization centers not from sites S_A but only from S_B.

(2) Initiation (formation of polymerization centers).

$$\text{Case A;} \quad S + (AlR_2Y)_2 \rightleftarrows S \cdot (AlR_2Y)_2 \qquad [30]$$

$$S \cdot (AlR_2Y)_2 + M \xrightarrow{\ \bar{k}_1\ } P^* \qquad [31]$$

$$\text{Case B;} \quad S + M \longrightarrow S \cdot M \qquad [32]$$

$$S \cdot M + (AlR_2Y)_2 \xrightarrow{\ \bar{k}_2\ } P^* \qquad [33]$$

where S is S_A or S_B in the case of Y = R, while S represents only S_B in Y = X.

(3) Propagation

$$P^* + M \xrightarrow{\bar{k}_P} P^* \qquad\qquad [34]$$

(4) Deactivation

$$P^*_A \xrightarrow{k_3} \text{dead center} \qquad\qquad [35]$$

$$P^*_B + M \xrightarrow{k_4} S_B \cdot M \text{ (in the absence of organo-metallic compound)} \qquad [36]$$

(5) Substitution of organomatallic compound

$$P^*(AlR_2Y) + AlEtCl_2 \longrightarrow \text{inactive center}$$

$$\text{wash} + AlR_2Y$$

$$\longrightarrow P^*(AlR_2Y) \qquad [37]$$

$$P^*_A(AlR_3) + AlEt_2X \longrightarrow \text{dead center} \qquad [38]$$

$$P^*_B(AlR_3) + AlEt_2X \longrightarrow P^*_B(AlEt_2X)$$

$$P^*_B(AlEt_2I \text{ or } AlEt_2Br) + AlEt_2Cl \longrightarrow$$

$$P^*_B(AlEt_2Cl \text{ for activity, } AlEt_2I \text{ or } AlEt_2Br \text{ for stereospecificity}) \quad [39]$$

$$P^*_B(AlEt_2Cl) + AlEt_2I \text{ or } AlEt_2Br \longrightarrow$$

$$P^*_B(AlEt_2I \text{ or } AlEt_2Br \text{ for both activity and stereospecificity}) \qquad [40]$$

$$P^*_B(AlEt_2X) + AlR_3 \longrightarrow P^*(AlR_3 \text{ for activity}) \qquad [41]$$

(6) Structures and reactions of polymerization centers. From the experiment of substitution of organometallic compound, we assume the existence of organometallic compound in a polymerization center. If the organometallic compound in the polymerization center P^*_B is in a coordinated form, S_B must have two vacant sites as below. Assuming an unstable Ti(III), which is not lattice Ti, we postulate the following structure for S_A:

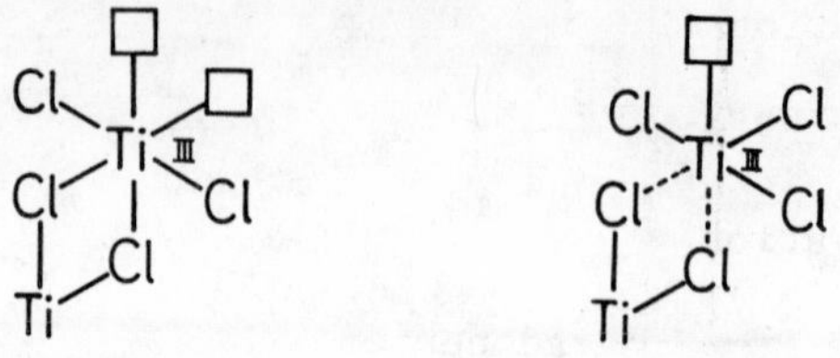

(S_B)   (S_A)

where the dotted line indicates dative bonds.

(i) Initiation

Case B

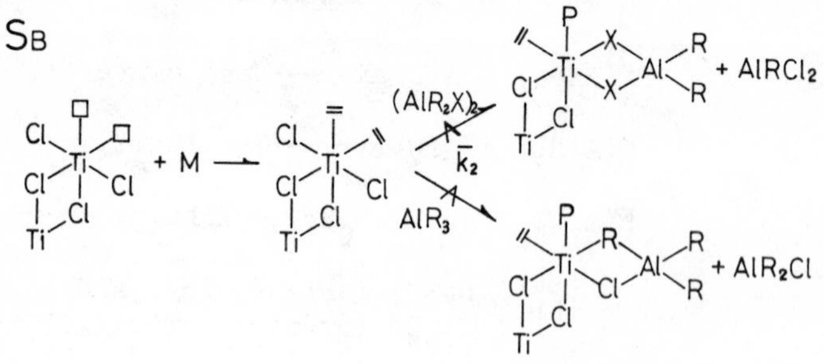

S_B

[42]

(P_B^*)

S_A

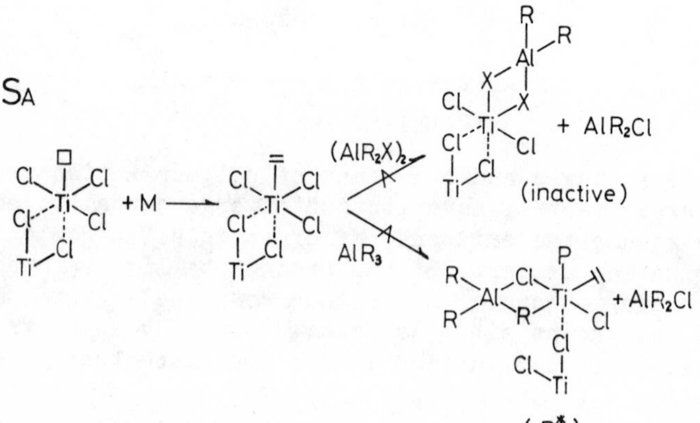

[43]

(P_A^*)

(ii) Propagation occurs by the collision of a monomer to the polymerization center, of which vacant site was occupied by a monomer.

(iii) Substitution of organometal.

$$
\begin{array}{c}
\text{P} \\
\diagup\!\!\!/\ \big|\ \diagup\!\text{I} \\
\underset{\text{Cl}}{\text{Ti}}\!\diagdown\!\underset{\text{I}}{\big|}\!\diagup\!\!\text{Al}\!\diagdown\!\overset{\text{Et}}{\underset{\text{Et}}{}}
\end{array}
+ (\,\text{AlEt}_2\text{Cl}\,)_2
$$

$$[44]$$

$$
\longrightarrow \quad
\begin{array}{c}
\text{P} \\
\diagup\!\!\!/\ \big|\ \diagup\!\text{I} \\
\underset{\text{Cl}}{\text{Ti}}\!\diagdown\!\underset{\text{Cl}}{\big|}\!\diagup\!\!\text{Al}\!\diagdown\!\overset{\text{Et}}{\underset{\text{Et}}{}}
\end{array}
+ \text{AlEt}_2\text{Cl}\cdot\text{AlEt}_2\text{I}
$$

(iv) Deactivation of washed catalyst by monomer.

$$
\begin{array}{c}
\text{P} \\
\diagup\!\!\!/\ \big|\ \diagup\!\text{X} \\
\underset{\text{Cl}}{\text{Ti}}\!\diagdown\!\underset{\text{X}}{\big|}\!\diagup\!\!\text{Al}\!\diagdown\!\overset{\text{R}}{\underset{\text{R}}{}}
\end{array}
+ \text{M} \longrightarrow
\begin{array}{c}
\text{P} \\
\diagup\!\!\!/\ \big|\ \diagup\!\!\!/ \\
\underset{\text{Cl}}{\text{Ti}}\!\diagdown\!\underset{}{\big|}\!\text{X}
\end{array}
+ \text{AlR}_2\text{X} \qquad [45]
$$

4) *GASEOUS POLYMERIZATION WITHOUT SOLVENT*

As noted before, the use of TiCl$_3$ aged with concentrated solution of AlEt$_3$ resulted in both lowering of the maximum rate of polymerization and increase of the isotacticity of produced polymers. This experimental result suggests a possibility that the polymerization with TiCl$_3$ and pure AlEt$_3$ in the absence of solvent gives highly isotactic polymers. This possibility was examined by the present author and coworkers(17,18). The kinetic curves of the propylene polymerization with TiCl$_3$ and pure AlEt$_3$ and AlEt$_3$-*n*-heptane are shown in Fig. 13. As expected, the maximum rate is depressed and the isotacticity of produced polymers is higher than that

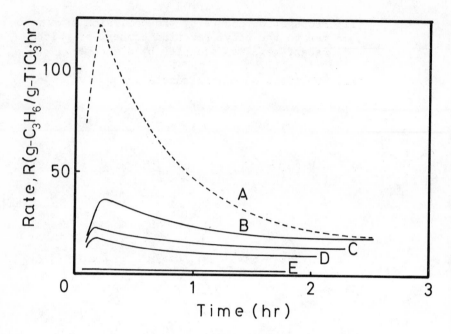

Fig. 13. Kinetic curves of slurry polymerization and gaseous propylene polymerizations with TiCl₃-AlEt₃ catalyst $P_{C_3H_6}$ = 76cmHg, 41°C. (A) slurry polymerization, TiCl₃ = 16mmol/l, [Al]/[Ti] = 1 n-heptane 200ml; (B) gaseous polymerization, [Al]/[Ti] = 2.5; (C) [Al]/[Ti] = 1.0; (D) [Al]/[Ti] = 0.7; (E) [Al]/[Ti] = 0.2, at TiCl₃ = 1.5 mmol.

of the usual slurry polymerization as shown in Table 3. The stationary rate of the polymerization could be represented by

$$R^G_\infty = R^L \frac{K_G P(A)}{1+K_G P(A)} \qquad [46]$$

where R^L_∞ is the stationary rate of the usual slurry polymerization and (A) is the amount of pure AlEt₃ added. Using a 1000ml glass reactor with a paddle stirrer, we have carried out the polymerization with TiCl₃-pure AlEt₃. It has been confirmed that the catalytic activity does not suffer by any physical process and remains unchanged long time, as shown in Fig. 14. The properties of produced polymers are summarized in Table 3.

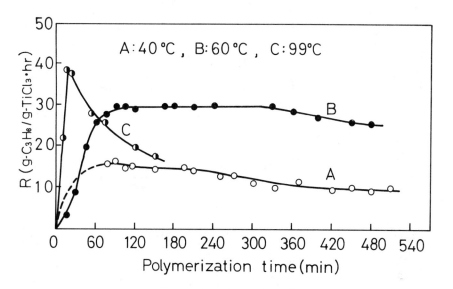

Fig. 14. Kinetic curves of gaseous propylene polymerization with TiCl₃ and AlEt₃. TiCl₃ = 1.5 mmol, [Al]/[Ti] = 2.5, P_C₃H₆ = 76 cmHg.

TABLE 3.

Isotacticities of polypropylene produced in gaseous polymerizations with TiCl₃-AlEt₃ catalyst at 41°C, 76 cmHg.

TiCl₃ (mmole)	AlEt₃ (mmole)	n-heptane (ml)	stationary rate (g-C₃H₆/g-TiCl₃/hr)	isotacticity[a,b] (%)	$\overline{M}_v$[a] (x10⁻⁴)
1.5	0.3	0	2.5	88.0	30.5
1.5	0.7	0	7.5	87.5	25.0
1.5	1.5	0	13.0	88.0	24.8
1.5	3.9	0	17.0	87.5	24.0
1.5	6.0	0	16.5	89.0	22.0
1.5	9.0	0	17.0	88.0	24.0
3.2	3.2	200	17.0	81.5	30.0
3.2	6.4	200	17.0	81.0	--

[a]The values obtained from total polymers produced at 90 min. [b]Isotactic fraction based on infra-red method.

LITERATURE CITED

(1) T. Keii, and Y. Doi, *Nature*, *242*, 14 (1973).
(2) T. Keii, Y. Doi, and H. Kobayashi, *J. Polym. Sci.*, *A-1*, *11*, 1881 (1973).
(3) Y. Doi, T. Hashimoto, and T. Keii, unpublished paper.
(4) T. Keii, "Kinetics of Ziegler-Natta Polymerization", Kodansha-Chapman Hall (1972). p. 15.
(5) These formulae were obtained with the polymerization with $TiCl_3$-$AlEt_3$, T. Keii, K. Soga, and N. Saiki, *J. Polym. Sci.*, *C16*, 1507 (1967).
(6) These have been confirmed with $TiCl_3$-$AlEt_2X$, T. Kohara, Y. Doi, and T. Keii, unpublished paper.
(7) T. Keii, M. Taira, and T. Takagi, presented at the Annual Meeting of Japan Chemical Soc. (1961). T. Keii et al., *Shokubai (Catalyst)* 5, 243 (1963).
(8) T. Keii, *Nature*, *196*, 160 (1962).
(9) A. K. Ingberman, I. J. Levine, and R. J. Turbett, *J. Polym. Sci. A1*, *4*, 2781 (1966).
(10) T. Keii, and T. Akiyama, *Nature*, *203*, 79 (1964).
(11) G. Natta, and I. Pasquon, *Advances in Catalysis*, *11*, 1 (1959).
(12) J. Ambrotz, P. Osecky, J. Mejzlik, and O. Hamerik, *J. Polym. Sci.*, *C16*, 42 (1967).
(13) T. Keii, T. Suzuki, and K. Soga, *Asahi Glass Kogyogijitsu Shoreikai Report (Japanese)*, *12*, 155 (1966), quoted fully in ref. 4, p. 45.
(14) T. Keii, and Y. Doi, *Kagaku no Ryoiki*, *26*, 793 (1972), in ref. 4, p. 101.
(15) D. F. Hoeg, and S. Liebman, Belq. Pat. 589260 (1960), quoted in D. O. Jordan, "The Stereochemistry of Macromolecules (ed. A. D. Ketley)", Vol. 1, Dekker, (1967), p. 31.
(16) A. Kojima, and T. Keii, unpublished paper, quoted in ref. 4, p. 203.
(17) S. Kanetaka, T. Takagi, and T. Keii, *Kogyo Kagaku Zasshi*, *67*, 1436 (1964), *69*, 695 (1966), *70*, 1568 (1967).
(18) Y. Doi, I. Okura, and T. Keii, *Chemistry Letters*, 327 (1972). Y. Doi, Y. Yoshimoto, and T. Keii, *Nippon Kagaku Zasshi*, 459 (1972). Y. Doi, H. Kobayashi, and T. Keii, *Nippon Kagaku Zasshi*, 1089 (1973). Y. Doi, Y. Hattori, and T. Keii, *Nippon Kagaku Zasshi*, 1636 (1973).

Chain Transfer in Ziegler Type Polymerization of Ethylene

G. HENRICI-OLIVÉ and S. OLIVÉ

Monsanto Research S.A.
8050 Zürich, Switzerland

1) INTRODUCTION

In general, the polymerization of ethylene with Ziegler catalysts is characterized by the absence of chain termination reactions, the length of the molecules being limited by chain transfer. In the absence of additives, the molecular weight determining reaction is the β-hydrogen abstraction (M = metal):

$$M-CH_2-CH_2-CH_2-CH_2\!\!\sim\!\!\sim \longrightarrow M-H + CH_2=CH-CH_2-CH_2\!\!\sim\!\!\sim \quad [1]$$

$$\qquad\qquad\qquad\qquad\qquad\qquad \underset{+\ C_2H_4}{\big\downarrow} \longrightarrow M-CH_2-CH_3$$

The relative frequency of the chain transfer reaction(1) depends upon the electron affinity of the transition metal center which may be influenced deliberately by suitable ligands(1,2). Several cases of soluble Ziegler type catalysts have been found recently, where the relative frequency of the β-hydrogen abstraction has been pushed far enough to obtain oligomers instead of the high molecular weight polymers originally aimed at (3-10). Of particular interest are those oligomerization catalysts which ensure linear chain growth; they not only produce the technically interesting linear α-olefins, but also permit a kinetic study, providing some insight into the detailed mechanism of the β-hydrogen abstraction reaction.

2) KINETIC EQUATIONS AND MOLECULAR WEIGHT DISTRIBUTION

In the absence of disturbing side reactions (in particular of the incorporation of the formed α-olefins into growing chains, leading to branching and vinylidenic or inner unsaturation(2,4), only chain propagation (rate r_p, rate constant

k_p) and chain transfer (r_{tr}, k_{tr}) have to be considered in the reaction scheme. The rate of chain propagation is given by

$$r_p = k_p [P^*][M] \qquad [2]$$

where $[P^*]$ and $[M]$ are the concentrations of active catalyst sites and of monomer respectively.

The rate of chain transfer may be either

$$r_{tr} = k_{tr}[P^*][M] \qquad [3a]$$

or

$$r_{tr} = k_{tr}[P^*] \qquad [3b]$$

depending on whether the monomer is involved in the rate-determining step of the chain transfer [cf. eq. 1]. Evidently, the overall kinetics depend upon this mechanistic feature.

Where eq. 3a is valid, monomer is used up not only by chain propagation but also in the transfer step. Hence, the overall rate of monomer consumption is

$$-d[M]/dt = (k_p + k_{tr})[P^*][M] \qquad [4a]$$

and the number-average degree of polymerization is

$$P_n = \frac{r_p + r_{tr}}{r_{tr}} = \frac{k_p}{k_{tr}} + 1 \qquad [5a]$$

If eq. 3b is valid:

$$-d[M]/dt = k_p[P^*][M] \qquad [4b]$$

$$P_n = r_p/r_{tr} = k_p[M]/k_{tr} \qquad [5b]$$

If the average molecular weight of the total oligomer is experimentally available, its monomer dependence allows one to decide which of the two reaction schemes (set a or set b) is applicable.

In either case the "most probable distribution of molecular weights" (Schulz-Flory distribution) is to be expected. The weight fraction m_p of product with degree of polymerization P is then given by(11,12):

$$m_p = \ln^2\alpha \cdot P \cdot \alpha^P \qquad [6]$$

or

$$\log(m_p/P) = \log(\ln^2\alpha) + P(\log\alpha) \qquad [7]$$

where the probability of chain growth, α, is given by eq. 8a or 8b, again depending on whether or not the monomer takes part in the rate determining step of reaction 1:

$$\alpha = r_p/(r_p + r_{tr}) = [1 + (k_{tr}/k_p)]^{-1} \qquad [8a]$$

$$\alpha = r_p/(r_p + r_{tr}) = [1 + (k_{tr}/k_p[M])]^{-1} \qquad [8b]$$

According to eq. 7, a plot of $\log(m_p/P)$ versus P permits the determination of $\log \alpha$ as the slope of a straight line. Hence, in cases where chromatographic fractionation data are available, α may be determined graphically. This is also true if low-molecular-weight material has been lost during working up, provided that the portion of the experimentally reliable fraction, say $C_{10}-C_{24}$, gives a linear plot.

An evaluation of the molecular-weight distribution (or parts thereof) thus provides an independent way of distinguishing between the two proposed reaction schemes. Once this problem is settled, the individual rate constants k_p and k_{tr} become accessible from the data on overall rate and molecular weight (eqs. 4 and 5) or from data on overall rate and fractionation (eqs. 4 and 8). The a or b set is used as it is appropriate but, of course, the concentration of active sites [P*] and concentration of monomer [M] must be known with sufficient reliability.

From a practical point of view it is often interesting to know which fraction of the total conversion can be expected to have chain lengths in a certain range, say C_6-C_{16}. This piece of information may easily be obtained with the aid of eq. 6 as

$$m_p(C_6-C_{16}) = \sum_{P=3}^{8} \ln^2\alpha \cdot P \cdot \alpha^P \qquad [9]$$

Eq. 9 gives this weight fraction as a function of α. This function (multiplied by 100 to give percentage) is represented in Fig. 1. The curve goes through a maximum which indicates the highest attainable fraction of C_6-C_{16}, as well as the α-value which must be aimed at in order to obtain these results. Fig. 1 includes also the functions

$$m_p(C_2+C_4) = \sum_{P=1}^{2} \ln^2\alpha \cdot P \cdot \alpha^P \qquad [10]$$

and

$$m_p(>C_{16}) = 1 - \sum_{P=1}^{16} \ln^2\alpha \cdot P \cdot \alpha^P \qquad [11]$$

The molecular weight distribution, eq. 6, takes into account also those molecules which have left the center with degree of polymerization P=1. Although these ethylene molecules do not contribute to the experimental overall conversion, they are also the product of catalytic steps. For practical use it might be more convenient to refer the weight fraction of olefin in a certain range to the total amount of ethylene converted to higher homologues. Mathematically that means to multiply Q with a factor f, given by

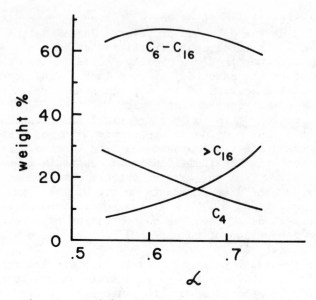

Fig. 1. Weight fractions of C_6-C_{16}, of $<C_6$ and of $>C_{16}$ as a function of α; calculated from eqs. 9–11.

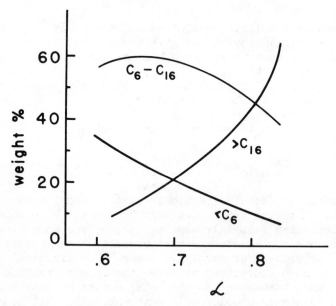

Fig. 2. Weight fractions, referred to experimental conversion (see text).

$$f = \frac{\sum\limits_{P=1}^{\infty} m_p}{\sum\limits_{P=2}^{\infty} m_p} = \frac{1}{1 - m_1} = \frac{1}{1 - \alpha \ln^2 \alpha} \qquad [12]$$

where Q is the original weight fractions for P⩾2, as calculated from eq. 6. The result is shown in Fig. 2. This graph shows that an oligomer of the Schulz-Flory type can have a maximum of 67.5% (by weight) of molecules in the technically interesting range C_6-C_{16}, provided the probability of chain propagation is experimentally adjusted at α = 0.62.

3) *EVALUATION OF EXPERIMENTAL DATA*

The catalyst system $(C_2H_5O)_3TiCl/C_2H_5AlCl_2$ in toluene has been shown to produce essentially linear α-olefins at temperatures below 0° and at an ethylene pressure of 12 atm, as long as the conversion is maintained below 10-15 moles of ethylene per liter of catalyst solution(2,4). Hence, the kinetic equations of the preceding section are in principle applicable to the oligomerization of ethylene with this catalyst.

Experimental data which allow one to decide between eqs. 5a and 5b are given in Table 1. It follows from the table that chain termination may be neglected (runs 1 and 2); the overall rate is proportional to the monomer concentration (runs 5-7); the molecular weight M_n depends neither on the monomer concentration nor on the catalyst concentration, but only on the temperature. Particularly, the independence of M_n upon the monomer concentration eliminates eq. 5b. This important result may be corroborated by literature data, obtained with another Ziegler type catalyst, $Zr(OC_3H_7)_4/$ $(C_2H_5)_3Al_2Cl_3$, which also was reported to produce linear α-olefins(6). Since in this case gas chromatographic fractionation data are available, α may be determined graphically according to eq. 7. Fig. 3 shows that a representation of the experimentally reliable fractions, C_{10}-C_{24}, gives a linear plot, (parts of the fractions <C_{10} might have been lost in the working up procedure). The monomer concentration varies by a factor of 4.5 within the three runs represented. The close similarity of the data in the C_{10}-C_{24} range excludes a monomer dependence of α according to eq. 8b, thus confirming the validity of the a-set of equations.

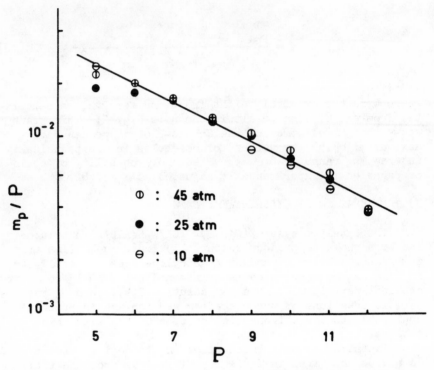

Fig. 3. Oligomerization of ethylene with $Zr(OC_3H_7)_4/(C_2H_5)_3$-Al_2Cl_3; molecular weight distribution at different ethylene pressure. Literature data(6) represented according to eq. 7. $T = 80°$; heptane ($\bullet$, $\oplus$) or toluene ($\ominus$).

The individual rate constants, k_p and k_{tr}, may now be determined from eqs. 4a and 5a in combination with the experimental data of Table 1, provided [P*] and [M] are known. Fig. 4 shows for the Ti/Al system treated in Table 1, that the overall rate depends on the ratio of aluminum alkyl to titanium compound, [Al]/[Ti], increasing in the range up to [Al]/[Ti] = 7, and then leveling off. At [Al]/[Ti]>7, the rate is proportional to the titanium concentration. These data permit the assumption that the active centers are formed from the two components in an equilibrium reaction and that at [Al]/[Ti]>7 essentially all titanium is present in the active form, i.e., [P*] = [Ti]$_{total}$. To evaluate the monomer concentration, literature data(13) have been extrapolated using Henry's law (-20°, [M] = 3.4 mole/l; -45°, [M] = 5.3 mole/l; at p = 12 atm). The individual rate constants, resulting from the runs at [Al]/[Ti] = 7 of Table 1, are given in Table 2 (System A).

TABLE 1.

Ethylene oligomerization with $(C_2H_5O)_3TiCl/C_2H_5AlCl_2$ in toluene.

Run No.	T, °C	$[Ti] \times 10^3$, mole/l	$P_{C_2H_4}$, atm.	t, hrs.	Conversion mole/l	M_n [b]
1	−45	20	12	1	4.8	90
2	−45	20	12	4	15.3	105
3	−20	10	12	1	9.0	138
4	−20	20	12	1	17.8	153
5[a]	−20	20	12	1	13.0	140
6[a]	−20	20	6	1	7.6	138
7[a]	−20	20	3	4	13.5	201

[a][Al]/[Ti] = 5; all other [Al]/[Ti] = 7.   [b]Corrected for loss of low-boilers(10).

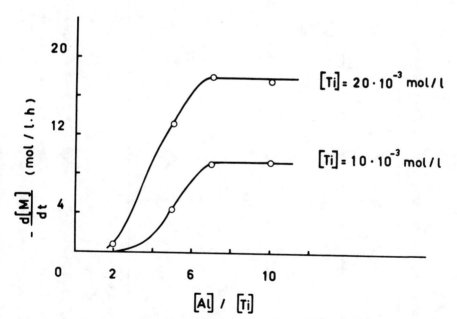

Fig. 4. Oligomerization of ethylene with $(C_2H_5O)_3TiCl/C_2H_5-AlCl_2$; dependence of the overall rate on the ratio [Al]/[Ti]. T = 20°; toluene; ethylene pressure 12 atm.(10).

TABLE 2.

Estimated rate constants of ethylene oligomerization with different catalyst systems.

Transition Metal	System*	T, °C	$P_{C_2H_4}$, atm.	α	k_p/k_{tr}	k_p (1/mol/sec)	k_{tr} (1/mol/sec)	Ref.
Ti	A	-45	12	0.74	2.8	8×10^{-3}	3×10^{-3}	10
Ti	A	-20	12	0.82	4.5	6×10^{-2}	1.3×10^{-2}	10
Ti	B	-20	11	0.75	2.9	7×10^{-2}	2×10^{-2}	5
Zr	C	+80	10	0.78	3.5	2×10^{0}	6×10^{-1}	6
Zr	D	+55	35	0.68	2.1	2×10^{0}	8×10^{-1}	7
Zr	E	+40	11	0.69	2.2	4×10^{0}	2×10^{0}	8
Ni	F	+75	38	0.70	2.3	2×10^{-4}	1×10^{-4}	14
Ni	G	+100	42	0.85	5.6	2×10^{-2}	3×10^{-3}	15

* A: $(C_2H_5O)_3TiCl/(C_2H_5)AlCl_2$/toluene. B: $TiCl_4$/t-BuOH/$(C_2H_5)_3Al_2Cl_3$/chlorobenzene.
C: $(C_3H_7O)_4Zr/(C_2H_5)_3Al_2Cl_3$/toluene. D: $(C_3H_7O)_4Zr/(C_2H_5)_3Al_2Cl_3$/n-heptane. E: Bz_4Zr/
$(C_2H_5)_3Al_2Cl_3$/toluene (Bz = benzyl). F: Bis-1,5-cyclooctadienenickel(0)/diphenylcarboxy-
methylphosphine/benzene. G: Bis-1,5-cyclooctadienenickel(0)/diphenyl(N,N-diphenylcarbamyl-
methyl)phosphine/benzene.

298

The data represented in Fig. 3 permit also an estimate of the individual rate constants, this time making use of eqs. 4a and 8a. A value of $\alpha = 0.78$ is obtained from the slope of the straight line. This gives $k_p/k_{tr} = 3.5$. For the particular case of ethylene pressure $= 10$ atm (solvent toluene), an overall rate of 5.4×10^{-3} mole/l/sec is reported, with $[Zr] = 2.3 \times 10^{-3}$ mole/l (7). Assuming that all Zr is active, and with $[M]_{40°}$, 10 atm ≈ 0.8 mole/l (13), eq. 4a gives $(k_p + k_{tr}) \approx 3$. The resulting individual constants are included in Table 2 (System C); these data and all others except those of system A represent lower limits because of the assumption $[Cat]_0 \equiv [P^*]$. It may reasonably be assumed that the kinetic scheme deduced for the systems A and C (ie., validity of the a-set of eqs. 3-5 and 8, is also applicable to other relevant systems. This permits then a rough estimate of the rate constants also in cases where the independence of P_n and α upon the monomer concentration has not explicitely been checked.

Langer(5) oligomerized ethylene with the catalyst system $(C_2H_5)_2AlCl/(C_2H_5)AlCl_2/TiCl_4/t$-BuOH and obtained linear α-olefins. For a molar ratio 6:6:1:1 of the four catalyst components, an ethylene pressure of $ca.$ 11 atm, and with chlorobenzene as solvent, an overall rate of 5.5×10^{-4} mole/l/sec and $M_n = 109$ are reported, for $[Ti] = 2 \times 10^{-3}$ mole/l at $-20°$. The rate constants estimated with eqs. 4a and 5a are given in Table 2 (system B).

In other cases gas chromatographic fractionation data are available permitting an evaluation with eqs. 4a and 8a. Longi, Greco and Rossi(7) reported the oligomerization of ethylene with $(C_3H_7O)_4Zr/(C_2H_5)_3Al_2Cl_3$ in heptane; the fractionation data represented according to eq. 7 give a straight line (Fig.5a), indicating that a Schulz-Flory distribution was obtained. The same is true for the system $Bz_4Zr/(C_2H_5)_3$-$AlCl_3$ in toluene (Bz = benzyl), which has been investigated by Attridge et $al.$ (Fig. 5b).

Two recent Shell patents report the oligomerization of ethylene to linear α-olefins with systems derived from bis-(cyclooctadiene)-nickel(0) and phosphines. These transition metal catalysts, although not of the Ziegler type (no aluminum alkyl), give also products having a Schulz-Flory distribution of molecular weights (Fig. 6). Estimated rate constants for the systems represented in Figs. 5 and 6 are included in Table 2.

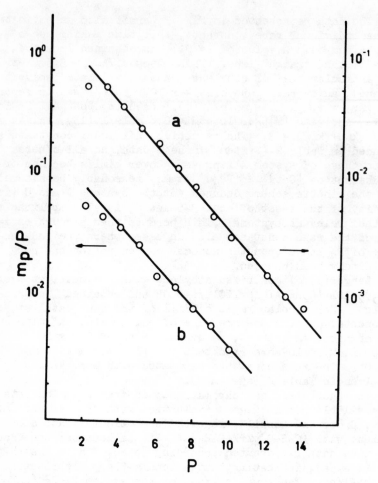

Fig. 5. Molecular weight distribution of oligoethylene obtained with zirconium catalysts. (a) System D (7); (b) System E (8) (cf. Table 2).

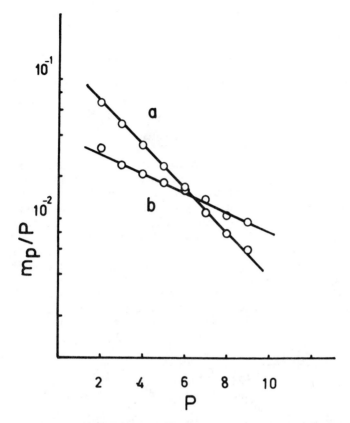

Fig. 6. Molecular weight distribution of oligo-ethylene obtained with nickel catalysts. (a) System F (14); (b) System G (15) (cf. Table 2).

In discussing the data of Table 2, it should be born in mind that α, and hence the ratio k_p/k_{tr} (eq. 8a) may be obtained from the experiments with a relatively high accuracy, whereas the individual constants k_p and k_{tr} are rough estimates.

For a given system, the temperature is the only variable for the adjustment of α, and hence for the optimization of the molecular weight range cf. eq. 8a and Fig. 2. It may be seen that the titanium catalysts require temperatures of <-20° in order to become interesting as oligomerization systems, whereas zirconium catalysts work satisfactorily at 40-80°. The great advantage of the latter is the reaction rate which lies some two orders of magnitudes higher than with titanium catalysts (rate constants k_p and k_{tr}). The apparent rate constants for the nickel systems, on the other hand, are extremely low even at high temperature. There might be some catalyst depletion by decomposition of the complex and precipitation of Ni° sponge.

4) THE MECHANISM OF THE β-HYDROGEN ABSTRACTION

The data for system A in Table 2 give the following temperature dependence for the individual rate constants:

$$\log k_p = 6.7 - 9200 / 4.57\ T \qquad [13]$$

$$\log k_{tr} = 3.9 - 6700 / 4.57\ T \qquad [14]$$

Although no more than rough estimates, the Arrhenius parameters of the rate constants merit some comments. The unusual temperature dependence of the molecular weight (smaller M_n with decreasing temperature, see Table 1) is traced back to the fact that $E_{tr} < E_p$. The low activation energy of the β-hydrogen abstraction is in fact amazing if compared, for instance, with that of the same reaction in aluminum alkyls, where an activation energy of 20-30 kcal/mole was reported (16). Moreover, the extremely low A-factor, as well as the occurrence of [M] in the rate law eq. 3a indicate a mechanism different from that leading to the aluminum hydride;

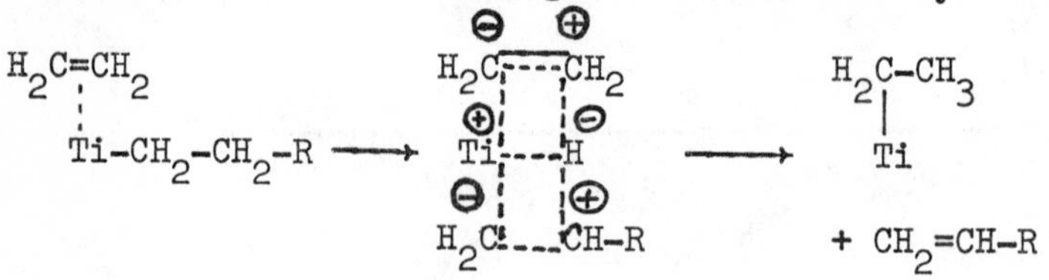

whereas the propagation step is assumed to proceed via a normal four center transition state:

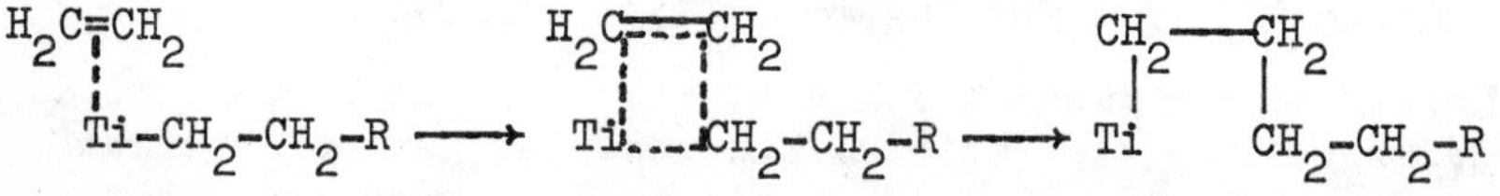

The mechanism suggested for the chain transfer reaction is distinctly different from that of other β-hydrogen abstraction reactions from transition metal alkyls. Several such reactions have been reported(17); they evidently proceed without the assistance of coordinated monomer.

LITERATURE CITED

(1) G. Henrici-Olive and S. Olive, *Angew. Chem.*, *83*, 121
     (1971); *Angew. Chem.*, *Int. Edit. 10*, 105 (1971).
(2) G. Henrici-Olive and S. Olive, *Chem. Ing. Techn. 43*,
     906 (1971).
(3) H. Bestian and K. Clauss, *Angew. Chem.*, *75*, 1068 (1963).
(4) G. Henrici-Olive and S. Olive, *Angew. Chem.*, *82*, 255
     (1970); *Angew. Chem.*, *Int. Edit. 9*, 243 (1970).
(5) A.W. Langer, *J. Macromolecular Sci. Chem. A4*, 775 (1970).
(6) Montecatini Edison s.p.A., Netherlands Patent Specn.
     7013193/1971.
(7) P. Longi, F. Greco, and U. Rossi, *Chimica E L'Industria
     (Milan) 55*, 253 (1973).
(8) C.J. Attridge, R. Jackson, S.J. Maddock, and D.T.
     Thompson, *Chem. Commun. 1973*, 132.
(9) G. Henrici-Olive, and S. Olive, *J. Polym. Sci. Polym.
     Lett.*, *12*, 39 (1974).
(10) G. Henrici-Olive, and S. Olive, *Adv. Polymer Sci.*, in
     press.
(11) G.V. Schulz, *Z. physik. Chemie B43*, 25 (1939).
(12) G. Henrici-Olive, and S. Olive: "Polymerization -
     Katalyse, Kinetik, Mechanismen", Verlag Chemie, Weinheim,
     Germany, 1969.
(13) J.A. Waters, G.A. Mortimer, and H.E. Clements, *J. Chem.
     Eng. Data 15*, 174 (1970).
(14) Shell, U.S. Pat. 3647906.
(15) Shell, U.S. Pat. 3647915.
(16) A.T. Cocks, and K.W. Egger, *J. Chem. Soc., Faraday I
     1972*, 423 (Vol. 68).
(17) P.S. Braterman and R.J. Cross, *J. Chem. Soc. Dalton*,
     1972, 657 and references therein.

Supported Ziegler-Natta Catalysts

JAMES C.W. CHIEN and J.T.T. HSIEH

*Department of Chemistry, Polymer Research Institute
and Materials Research Laboratories
University of Massachusetts
Amherst, Massachusetts 01002*

1) INTRODUCTION

Supported catalyst for olefin polymerization is of course not new. The Phillips catalyst, which is chromium supported on SiO_2/Al_2O_3, is used to produce much of the high density polyethylene in the world today. Unfortunately this catalyst cannot be used to polymerize propylene. The Ziegler-Natta catalyst, which is the preferred catalyst in the manufacturing of semicrystalline polypropylene, has several drawbacks. In this system the efficiency of utilization of Ti, i.e., grams of polymers produced per mmole of Ti, is quite low. This is because the basal faces of the α-TiCl$_3$ crystals which consist of only Cl atoms are inactive. Only those coordinatively unsaturated Ti atoms along the lateral edges provide active sites for polymerization(1,2). Consequently, the process requires costly purification steps to remove the catalyst residues. Secondly, there are probably several kinds of sites each with different polymerization activity and stereospecificity. The polymers obtained with Ziegler-Natta catalysts have broad molecular weight distributions and are sometimes difficult to process. Melt fracture and extrusion instability have often been encountered. These reasons have led to recent interests in the development of second generation Ziegler-Natta catalysts.

The working hypothesis adopted in our work is the following. First one selects an inert support which has well characterized surface functionalities. Transition metal compounds are impregnated on the support by chemical reactions with the surface functional groups. Upon activation the

*This work is supported by a grant GH 39111 from the National Science Foundation.

active center should then be chemically uniform and each capable of initiating polymerization. The stereospecificity can be controlled by the nature of the transition metal compounds and the support. In principle, such a catalyst system should give stereoregular polymer with high efficiency. Since the support is inert it may be possible to leave it in the polymer. This general approach seems to be shared by other scientists as well who are engaged in this line of research(3-9).

In this paper we will present our work on the characterization of surface hydroxyls on catalyst supports and the polymerization of olefins with various supported catalysts.

2) CHARACTERIZATION OF SURFACE HYDROXYLS WITH PARAMAGNETIC PROBE

2.1) Principle of the Method

The characterization of a catalyst support material begins with the determination of surface area. BET method is generally used and will not be elaborated here. The particular kinds of supports of interest to us are metal oxides and oxyhalides. The surfaces of these materials are rich in hydroxyl groups to which organotransition metal compounds and transition metal halides can be attached. The number of surface hydroxyls increases with the decrease of particle size and is also a function of thermal history. For these materials we need to know, in addition to the surface area, (1) number of surface hydroxyls per unit area (OH/100 $Å^2$), (2) types of surface hydroxyls, i.e., isolated, near neighbors and germinal, (3) geometric distribution of surface hydroxyls, and (4) state of the transition metal on the surface after it is impregnated.

The concentration of surface hydroxyls can be readily measured with several methods such as TGA, neutralization, D_2O exchange, and reactions with thionylchloride, CH_3MgI, and metal chlorides(10). Infrared spectroscopy has been used to differentiate hydrogen-bonded OH groups and free OH groups(11). From the stoichiometries of reactions with $SiCl_4$ and $AlCl_3$, Peri and Hensley(12) estimated the fraction of hydroxyl groups which are paired.

We have developed a paramagnetic probe method(13) which is capable of yielding useful information on all four points above. The technique involves the titration of surface hydroxyls with gaseous VCl_4 followed by the recording of the electron paramagnetic resonance (epr) spectra. The principles involved in this method are the following.

VCl_4 is a paramagnetic molecule with orbitally degenerate ground state. Its relaxation time is too short to permit epr observation even at $77°$ K(14). A broad epr signal can be seen at $4°$ K(15,16). However, when this degeneracy is lifted such as through replacement of one or more of the Cl atoms with alkoxy groups(14), the epr spectra of the vanadium atom can be observed even at room temperature because of greatly increased spin-lattice relaxation time.

By working with well characterized fumed silica, we found that several different types of surface hydroxyls can be distinguished by the vanadium epr spectra. With isolate hydroxyls, VCl_4 reacts and becomes attached to the surface via a single O-V bond (I). The reaction can be represented by

$$VCl_4 + \exists\text{-OH} \longrightarrow \underset{(I)}{\exists\text{-OVCl}_3} + HCl \qquad [1]$$

The epr spectrum of I is nearly isotropic: $g_o = 1.953 \pm 0.001$ and $A_o = 120 \pm 2$ G (Fig. 1). Reaction of VCl_4 with vicinal pair of hydroxyls, separated by a distance of about 3.2 Å,

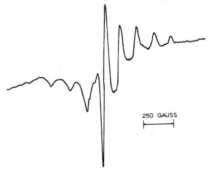

250 GAUSS

Fig. 1. Epr spectrum of VCl_4 on annealed Cab-O-Sil at $-195°$ 10-30% coverage.

forms species II,

$$VCl_4 + \begin{array}{c}\exists\text{-OH} \\ \exists\text{-OH}\end{array} \longrightarrow \begin{array}{c}\exists\text{-O} \\ \exists\text{-O}\end{array}\!\!\!\!>\!VCl_2 + 2HCl \qquad [2]$$

(II)

This species (on Cab-O-Sil) has pronounced axial anisotropy: $g_\perp = 2.0250$, $g_{||} = 1.9699$, $A_\perp = 79.3$ G, and $A_{||} = 189.9$ G (Fig. 2). Similar spectra were observed for VCl_4 on $\gamma\text{-Al}_2O_3$ which has close-packed hydroxyls. The epr spectra of VCl_4 on boehmite has rhombic characteristics: $g_1 = 1.905$, $g_2 = 1.974$, $g_3 = 1.986$, $A_1 = 192.6$ G, $A_2 = 161.1$ G and $A_3 = 82.3$ G (Fig. 3).

307

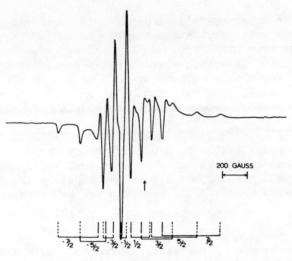

Fig. 2. Epr spectrum of 5% vanadium on rehydrated Cab-O-Sil.

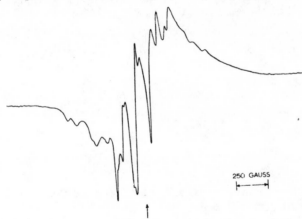

Fig. 3. Epr spectrum of VCl_4 on boehmite surface, 0.495 mmole of V g^{-1}.

This is probably a species like *II* except the ligand field of the vanadium is under further influence of adjacent polar groups. The possibility that the spectrum may be the sum of two different types of species *II* cannot be entirely excluded.

With increasing coverage of vanadium, the epr spectra become broadened and diminished in intensity. This is the result of exchange interaction, JS_1S_2, wher J is the exchange integral and the S's are the spin angular momentum. Using the model of small admixture of a charge-transfer excited state,

we estimated that when two vanadium atoms fixed on the surface is separated by a distance of about 8 Å the exchange
interaction would obliterate the hyperfine structure, i.e.,
$J \approx A$. The temperature dependence of line broadening can be
used to distinguish I from II and permits an estimate of
separation between these species. If only II are formed on
the surface, then the broadening process is temperature independent and is only a function of its concentration (fractional
coverage). This is because II is fixed in space. In contra-
distinction, the broadening phenomenon is quite dependent on
temperature for species I. The reason lies with the ability
of its vanadium to rotate about the single O-V bond. At
elevated temperatures, this rotation can bring two singly
attached vanadium (I), which are separated by as much as 12 Å,
sufficiently close to have $J \approx A$. Similarly, it is 10 Å for
effective exchange broadenings between a singly and a doubly
attached vanadium. At low temperatures, the interacting
species are frozen and immobilized; those which are separated
more than 8 Å contribute toward narrow epr signals, and others
give broad resonances.

Finally there are germinal hydroxyls which can react with
two molecules of VCl_4

$$[3]$$

$$(I') \qquad\qquad (III)$$

Species I' behaves just like I, but III has no observable epr
at any temperature. The vanadium atoms in III are so close to
one another that in addition to exchange interaction, dipolar
interaction and moderation of zero-field splitting render the
epr line-width excessively broad for detection.

2.2) Rehydrated Cab-O-Sil

Reflux of high temperature dehydrated Cab-O-Sil in water
produced a product with very reproducible amounts of 4.6 OH/
100 Å^2. It has been suggested(17) that the surface resembles
the [0001] face of β-tridymite and that the hydroxyls are
uniformly found at 5 Å from one another.

We found by the paramagnetic probe method that reaction
of VCl_4 with rehydrated Cab-O-Sil yielded $\sim$50% of species I
and I'. The results are inconsistent with a structure similar
to the [0001] face of β-tridymite, but the structure is rather
akin to the [001] face of β-cristobalite. The latter model
has 70% of the hydroxyls as vicinal pairs and the remainder

309

as germinal pairs both distributed randomly over the surface(12).

2.3) Annealed Cab-O-Sil

When Cab-O-Sil is heated at 700 ± 25° for 48 hours there is formed a product which has reproducible amounts of 1.4 OH/ 100 $\overset{\circ}{A}^2$. At even the lowest coverages, the epr signal are weak at room temperature but intense at 77° K. The intensity at 77° K decreases rapidly with increase of coverage above 50%. These results are consistent with a model of surface where only clusters of germinal OH pairs are present. Random dehydration of a surface such as the [001] face of β-cristo-balite(3) would reduce the germinal OH pairs to 1.4 OH/100 $\overset{\circ}{A}^2$. The conversion of the silanol groups, which are on the average 5 A apart before heating, requires the rotation of neighboring SiO$_4$ tetrahedra to reduce markedly the separation of the Si atoms and to bring the bridging oxygen atom into line to link the tetrahedra. This is not unlike the creation of elevation and depression in the earth topology. This kind of deforma-tion could produce islands of hydroxyls which were observed to remain after annealing.

2.4) Alumina

Based on its unit cell structure γ-Al$_2$O$_3$ has 17 OH/100 $\overset{\circ}{A}^2$ of surface. The average separation of the hydroxyls are par-ticularly favorable for reaction 2. This is born out by the paramagnetic probe study of fumed alumina (Alon)(15). This material has a surface of 100 m^2 g^{-1} and a nearly theoretical amount of hydroxyls (2.8 mmole g^{-1}). After reaction with VCl$_4$, its room temperature and 77° K epr spectra are identical in features at all coverages, they differ only in intensities in accord with the Curie dependency. The epr intensity increases to a maximum at about 10% coverage and decrease to nearly zero at greater than 40% coverage due to dipolar and exchange broadening.

When Alon is heated in water to 200° under pressure for two hours, the material is transformed to gibbsite. Para-magnetic probe results showed that the VCl$_4$ reacted to give II, the epr spectral parameters for II on either fumed alumina or gibbsite are: $g_\perp$ = 2.01, $g_{||}$ = 1.95, A$_\perp$ = 70 G, and A$_{||}$ = 167 G.

Heat treatment of Alon in water to 250° for six hours affords boehmite having both hydroxide and oxide ions in the surface. Reaction with VCl$_4$ led to species II' whose rhombic epr spectra were already described above.

Since the acidity of metal oxide can have profound effects on the activity of the material as a catalyst or a support, we examine three samples of CAMAG alumina to see whether paramagnetic probe would be capable of determining the acidity of surface hydroxyls. The physical properties of the aluminas are given in Table 1, they all have 155 m^2 g^{-1}

TABLE 1.

Physical properties of CAMAG aluminas

Sample	pH	Na_2O %	Contents Cl^-, meq/g	OH, mmole/g
Neutral alumina (507-C-1)	7.0±0.5		0.03	1.41
Basic alumina (5016-A-1)	9.5±0.2	0.15		1.17
Acidic alumina (504-C-1)	4.5±0.5		0.14	1.93

of BET surface area and 58 Å average diameter of micropores. The OH contents were measured with a DuPont 950 TGA apparatus. Their surface hydroxyl contents are only about one half of the Alon specimen. Reaction with VCl_4 yielded mainly type *II* products, the epr spectra for all the CAMAG samples are indistinguishable from those of Alon. In these materials the acidity and basicity are controlled by Cl^- and Na^+ substitution. However, the electron density differences are not transmitted to the vanadium atoms bond to the hydroxyls. All the spectra have identical *g* values and hyperfine anisotropy.

The lower surface hydroxyl contents of CAMAG aluminas do affect the paramagnetic probe results in another way. The epr intensities increase rapidly at low coverages, but leveled off at about 20% coverage. By comparison, the vanadium epr intensity on Alon increases with coverage to a maximum value at about 10% coverage. Further reaction causes rapid decreases in epr intensities(13). Therefore, about 20% of the surface hydroxyls in CAMAG has no neighboring hydroxyls. Its surface is quite irregular and is not that of a close-packed hydroxyl surface of Alon.

2.5) <u>Magnesium Hydroxylchloride</u>

VCl_4 has been reacted with $Mg(OH)Cl$ from 5 to 81% coverage. The epr spectra were recorded at 25° and –195°. A typical spectrum for a 20.4% coverage sample at –195° is shown in Fig. 4. The spectra at all other coverages and at either 25° or –195° are essentially the same as Fig. 1 except for intensity differences. The spectrum is typical for an axial system. The spectral parameters are: $g_{||}$ = 1.956, $g_\perp$ = 2.010, $A_{||}$ = 185 G, $A_\perp$ = 66.3 G. There are no broad signal underlying the spectrum with resolved hyperfine structures at all temperatures and all coverages of VCl_4.

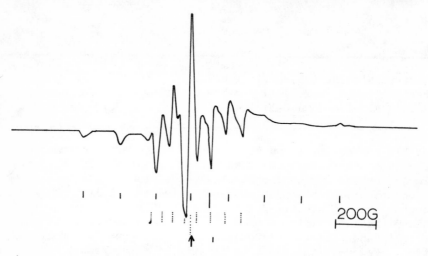

Fig. 4. *Epr spectrum of VCl_4 on $Mg(OH)Cl$, 22% coverage.*

Fig. 5 is a plot of epr intensities with monolayer coverage. The plot is linear at low coverage <40%; there is no temperature dependent broadening or intensity anomalies. Calibration with $VO(acac)_2$ solution of known concentration showed that all the vanadium in the surface contribute toward the epr spectrum. It can be said that all the VCl4 react to give type *II* products. Above 40% coverage the epr intensity begins to decrease with increasing coverage. VCl4 must react with surface hydroxyl pairs separated by <6 A at high coverages.

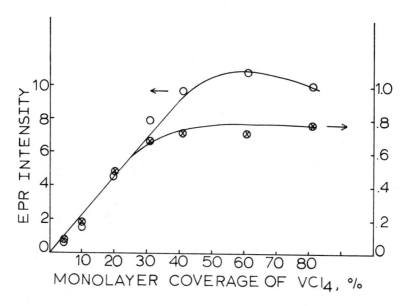

Fig. 5. *Variation of epr intensity with coverage of VCl$_4$ on Mg(OH)Cl.*

3) *STUDY OF TERMINATION REACTION ON SUPPORT SURFACES*

It was pointed out in Section 2.1 that the vanadium atoms in the type *I* and *III* species can approach other neighboring vanadium atoms by virtue of their rotational freedom. This has important consequence in limiting the chain length of polymerization. Inactivation of growing chains often result from reductive disproportionation, i.e.,

$$Cl_3TiR + Cl_3TiR' \rightarrow 2TiCl_3 + R_{-H} + R'_{+H} \quad [4]$$

This reaction has been shown to occur between $Cl_3TiC_2H_5$ and during polymerization initiated by $(\pi\text{-}C_5H_5)_2TiCl_2$ and aluminum alkyl halides(18). Similar reactions also have been reported for CH_3TiCl_3 with and without $TiCl_4$(19-22). A method has been developed by us(23) to study this important process.

The method is based on the reaction of surface hydroxyl group with $TiCl_4$. The attached Ti(IV) is subsequently alkylated at low temperature. Under conditions which Ti(IV) alkyls are stable in dilute solution, bimolecular reduction occurs when they are situated on adjacent surface sites.

The results on annealed Cab-O-Sil showed that all Ti(IV) alkyls undergo reduction at -80°C. All the Ti(IV) alkyls are singly attached based on the VCl_4 experiments. It is possible for alkyls situated as much as 10 Å apart to react with each other. On the other hand, the reduction was only 47% complete for rehydrated Cab-O-Sil because about half of the Ti(IV) alkyls are of type *II*. Bimolecular reduction of type *II* species is impeded by lack of freedom of rotation.

4. ETHYLENE POLYMERIZATION

Table 2 summarized the results of ethylene polymerization catalyzed by $TiCl_4$ supported on annealed Cab-O-Sil activated by $Al(i-Bu)_3$. These reactions are characterized by very fast initial rate of polymerization. This initial rate is difficult to measure accurately both because it is fast and also it decays rapidly. The rate of polymerization after 2 hours is about 10-30 gm polymer/mmole Ti/hr/atm. C_2H_4.

TABLE 2.

Ethylene polymerization with catalyst supported on annealed cabosil. *

[Ti] mM	$[Al(i-Bu)_3]$ mM	Temp °C	gm polymer/mmole Ti/ hr/atm C_2H_4 Initial	Final	Polymer yield gm/mmole Ti
0.1	0.2	50	230	10	410
0.1	0.6	50	400	20	390
0.3	0.6	50	100	30	450
0.07	0.28	50	310	20	450
0.1	0.6	80	90	20	300

*
The catalyst contains 0.3 mmoles of $TiCl_4$ per gram of annealed Cab-O-Sil it is activated by reacting the catalyst slurry at 7 mM Ti with 3.5 mM $Al(i-Bu)_3$ for 1 hour at 0°C, polymerization time 180 min.; $p_{C_2H_4} = 3$ atm.

Paramagnetic probe study showed that annealed Cab-O-Sil has clusters of geminal hydroxyls. Reaction with $TiCl_4$ and alkylation creates many active sites for polymerization. But they undergo rapid bimolecular reduction as shown by epr results. The decay of polymerization activity continues until each active site becomes well separated from another.

Rehydrated Cab-O-Sil is just as disappointing as annealed Cab-O-Sil as a catalyst support (Table 3). Because of greater

Table 3.
Ethylene polymerization with catalyst supported on rehydrated cabosil. *

[Ti] mM	[Al(*i*-Bu)$_3$] mM	Polymerization rate gm polymer/mmole Ti/hr/atm C_2H_4		Polymer yield gm/mmole Ti
		Initial	Final	
0.3	0.3	40	15	220
0.3	0.6	90	30	380
0.3	0.6	67	30	380
0.3	0.9	70	30	330
0.3	0.6	62	25	335
0.3	0.6	80	20	200

* The catalyst contains 0.6 mmoles of $TiCl_4$ per gram of rehydrated Cab-O-Sil, polymerization at 50° for 180 min; $P_{C_2H_4}$ = 3 atm.

surface hydroxyl content, it is possible to incorporate more $TiCl_4$ per gram of rehydrated Cab-O-Sil. But the initial polymerization activity is much lower than the catalyst based on annealed Cab-O-Sil. The results suggest that type *II* Ti is less active than type *I*. The decay of polymerization activity of rehydrated Cab-O-Sil catalyst is not as drastic as that of annealed Cab-O-Sil. After 180 min. the two systems have comparable activities. Fig. 6 illustrates the kinetics of polymerization for rehydrated Cab-O-Sil.

A number of aluminium alkyl activators were used with the rehydrated Cab-O-Sil catalyst. The trialkyl aluminum is superior to dialkyl aluminum chlorides with monoalkyl aluminum dichlorides being the most inferior activator (Table 4).

The polymerization of ethylene by $TiCl_4/Al_2O_3$ behaves much like that of the Cab-O-Sil systems. Fig. 7 shows the course of a typical polymerization.

5) *POLYMERIZATION OF PROPYLENE*

Through the work of Pino on stereoselective and stereoelective polymerizations, it has been established that it is the metal atom which exercises steric control of monomer insertion in stereoregular polymerization of α-olefins. In

315

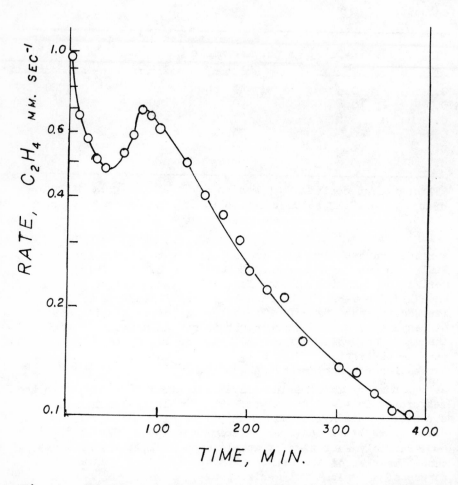

Fig. 6. *Ethylene polymerization by* $TiCl_4/Cab-O-Sil/AlEt_2Cl$ *catalyst at 50°.*

TABLE 4.

Ethylene polymerization with catalyst supported on rehydrated Cabosil[a] with a number of aluminum alkyls.

[Ti] mM	Aluminum Alkyls	Temp. °C	Time, min	Polymerization Rate, gm/mmole Ti/hr/atm Initial	Final	Polymer yield, gm/mmole Ti
0.3	$(CH_3)_2AlCl$, 0.6mM	50	200	20	8	110
0.6	$(i\text{-}Bu)AlCl_2$, 0.6mM	50	180	4	4	30
0.6	$(i\text{-}Bu)_2AlCl$, 0.6mM	50	220	27	7	140
0.6	$(C_2H_5)_2AlCl$, 0.6mM	50	200	48	20	310
0.6	$(C_2H_5)_2AlCl$, 0.6mM	80	200	60	0	150
0.3	$Al(i\text{-}Bu)_3$, 0.9mM	50	200	70	30	330

[a]The catalyst contains 0.6 mmoles of $TiCl_4$ per gram of rehydrated cabosil. $P_{C_2H_4}$ = 3 atm., polymerization time = 180 min.

317

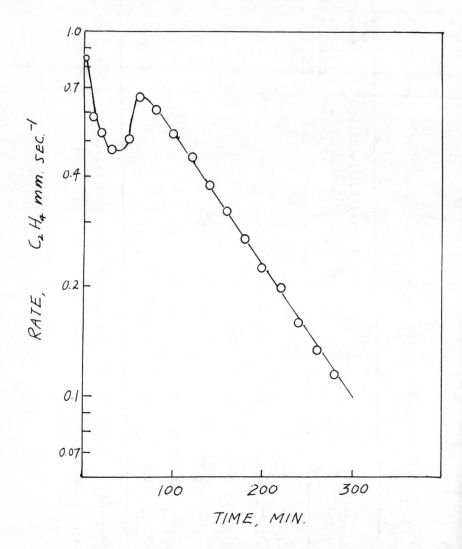

Fig. 7. Ethylene polymerization by $TiCl_4/Al_2O_3/AlEt_2Cl$ catalyst at 50°.

developing a supported catalyst for stereospecific polymerization of propylene we require transition metal after alkylation to become a chiral catalytic center capable of distinguishing the two antipods of the monomer. Furthermore, the transition metal should be doubly attached to the support so that its chirality is maintained during polymerization. Finally, electron donors are used to modify stereoselectivity of the catalysts.

5.1) Catalyst Screening Experiments

A number of transition metal alkyls and aryls were screened as catalyst with a variety of supports. $AlEt_2Cl$ was used as the activator. Polymerizations were carried out at 45-50° and 1.5 atm. propylene pressure with n-heptane as the diluent. The results are summarized in Table 5.

Zeolite, TiO_2, $Ca[OH]_2$ and parasym are all very poor supports. Al_2O_3, charcoal, and $Mg(OH)Cl$ are relatively good support materials. Chromocene supported on silical alumina has been reported to be a good catalyst for ethylene polymerization(6-8). In our work, hafnocene and titanocene on $Mg(OH)Cl$ have low activity for polymerization, which gave no high molecular-weight products.

$(\pi-C_5H_5)TiMe_3/Mg(OH)Cl$ has high polymerization activity, about 90% of the polypropylene is insoluble in boiling heptane. If the titanium is doubly attached to the surface then, the active site M is a chiral center where M is the monomer.

$$\equiv \begin{array}{c} - O \\ - O \end{array} \hspace{-4pt} > \hspace{-2pt} \underset{\underset{Al}{|}}{Ti} \hspace{-2pt} < \hspace{-4pt} \begin{array}{c} (\pi-C_5H_5) \\ R \end{array}$$

5.2) Various Ti alkyls as Catalyst

Four different Ti-alkyls were investigated as catalysts. On $Mg(OH)Cl$ support, tetrabenzyl titanium (B_4Ti) gave good rates of polymerization and fairly high stereospecificity (Table 6). Tribenzyltitanium chloride (B_3TiCl) is somewhat more active than B_4Ti. The chloride ligand usually has activating effect on coordinated anionic polymerization. The stereospecificity of B_3TiCl, is however, somewhat less than B_4Ti. One would have expected an increased rate with p-methylbenzyl derivative because of increased electron density over benzyl derivative. Any increased rate of initiation is, however, an insignificant contribution to the rate of polymerization. Stereospecificity of $(C_8H_9)_4Ti$ is greater on SiO_2 support as compared to $Mg(OH)Cl$. However, the activity is too low with SiO_2 for it to be of any interest as a viable catalyst.

319

TABLE 5.

Propylene polymerization

	Catalyst System						
Designation	Transition metal	mmole/gm support	Support[a]	A/M[b]	Electron donor	Polymer-ization rate[c]	% Stereoregular polymer
21	B₄Ti[d]	0.99	Al₂O₃	3.7	Ether	0.45	58
22	B₄Ti	0.44	charcoal	8	Ether	0.25	71.5
23	B₄Ti	0.30	Zeolite	6.4	Ether	0	0
24	B₄Ti	0.14	Ti₂O	22	Ether	0.12	n.d.[d]
25	B₄Ti	0.14	Ca(OH)₂	4.4	Ether	0.14	n.d.
26	B₄Ti	0.14	Parasym SMM-100	0.14	Ether	0	0
27	B₃TiCl[f]	0.14	Ca(OH)₂	21	Ether	0.07	n.d.
29	B₄Zr[g]	0.25	Mg(OH)Cl	21	Ether	0.08	>90
31	(π-C₅H₅)₂Hf	0.12	Mg(OH)Cl	48	THF	0.34	All low M.W.
32	(π-C₅H₅)₂Ti	0.16	Mg(OH)Cl	18	THF	0.11	All low M.W.
42	(π-C₅H₅)TiMe₃	0.07	Mg(OH)Cl	26	---	1.55	89.5
28	(π-C₅H₅)₂ZrMe	0.15	Mg(OH)Cl	20	Ether	0.03	n.d.
37	α-TiCl₃(AA)			1.4	---	0.53	95
38	α-TiCl₃(AA)			5.3	---	0.97	97

[a] The surface areas in m²/g for the supports are: Al₂O₃(100); Calgon charcoal (1200); Mg(OH)Cl(5). [b] ALEt₂Cl: transition metal ratio. [c] g polymer/mmole transition metal/hr/atm. propylene. [d] Tetra-benzylitanium. [e] not determined. [f] Tribenzylitanium chloride. [g] Tetrabenzylzirconium.

TABLE 6.

Propylene polymerization catalyzed by various Ti alkyls.[a]

Designation	Ti Alkyl	mmole Ti/ gm support	Al/Ti	Polymerization rate[b]	% Stereoregular polymer
6[c]	B_4Ti	0.074	25	1.12	77
11[c]	B_3TiCl	0.073	12.5	2.16	67
12	B_3TiCl	0.18	10	1.18	56
14	B_3TiCl	0.13	8	0.8	60
13	B_3TiCl	0.074	25	2.89	68
33	$(C_8H_9)_4Ti$[d]	0.08	22	1.04	65
35	$(C_8H_9)_4Ti/SiO_2$	0.60	22	0.10	82
34	$(C_8H_9)_3TiCl$	0.08	22	1.17	67

[a]On $Mg(OH)Cl$ support except run 35 which was on Cal-O-Sil support; [b] g polymer/mmole Ti/hr/atm propylene; [c]catalyst prepared in the presence of monomer, all others under argon; [d]tetra(p–methyl benzyl) titanium.

5.3) Effect of Electron Donor

Electron donor additives are known sometimes to have a marked effect on coordination polymerization. We briefly examined the effect of diethyl ether and tri(n-butyl)amine on propylene polymerization (Table 6). The effect of ether on B$_4$Ti catalyst is largely detrimental. It reduces the rate of polymerization as well as lowers the stereoregularity of the polymer. Tri-t-butylamine does not have significant effect on either the rate or the stereospecificity of the polymerization.

Ether is definitely beneficial to the B$_3$TiCl catalyst. Even though it reduces the rate of polymerization by about four-fold, the polymer formed is all insoluble in boiling heptane. The effect of ether may be two-fold. It may be chemisorbed on some of the very acidic surface sites to prevent the reaction of Ti alkyl with these sites. If these sites have low stereospecificity, poisoning of these could increase overall stereospecificity of the catalyst. Secondly, either could complex with Ti to improve steric control with some reduction in the rate of polymerization. The two effects are not mutually exclusive and may in fact be both operative.

In Table 7 it is also seen that excess of AlEt$_2$Cl tends to lower stereospecificity of the catalyst. This effect is a fairly general one also seen with other catalysts not reported here.

5.4) Kinetics of polymerization

Unlike the ethylene polymerization described above which has rapidly decaying rates, the propylene polymerization catalysts studied here have relatively constant rate of polymerization. The polymerization of B$_4$Ti/Mg(OH)Cl catalyst slowly increases to a constant rate. Fig. 8 shows a comparison of the results with those of an α-TiCl catalyst. The constant rate of polymerization suggests that the catalytic centers have long life-times. The polymer chains are probably terminated by chain-transfer processes such as transfer to monomer, to AlEt$_2$Cl, or β-hydrogen elimination leading to titanium hydride and polymer.

For the more active catalysts, such as B$_4$Ti/Mg(OH)Cl (Fig. 9), the rate increased to a maximum which declined slightly to a constant level. Both of the above kinetic behaviors have been seen with α-TiCl catalyzed propylene polymerization(24-28).

TABLE 7.

Effect of electron donors on propylene polymerization.

Designation	Ti Alkyl	mmole Ti/ gm support	Al/Ti	Electron Donor	Polymerization rate[b]	% Stereoregular Polymer
3	B_4Ti	0.23	3.8	ether	0.13	69
4	B_4Ti	0.19	6.4	ether	0.43	71
2	B_4Ti	0.13	21	ether	0.52	50
6	B_4Ti	0.07	25	---	1.12	77
9	B_3TiCl	0.14	10	ether	0.27	100
10	B_3TiCl	0.13	21	ether	0.28	70
12	B_3TiCl	0.17	10	---	1.18	56
14	B_3TiCl	0.13	8	---	0.8	60
8	B_4Ti	0.073	25	$(n-Bu)_3N$	0.87	76
15	B_3TiCl	0.074	25	$(n-Bu)_3N$	1.37	63

[a]The support is Mg(OH)Cl; [b]g polymer/mmole Ti/hr/atm. propylene.

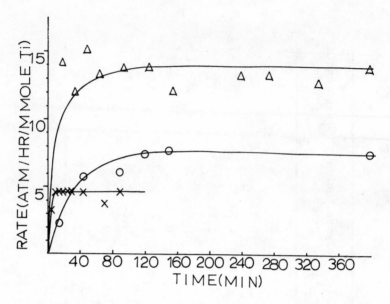

Fig. 8. Propylene polymerization at 50°. Δ B₄Ti/Mg(OH)Cl/
AlEt₂Cl/(n-Bu)₃N; O B₄Ti/Mg(OH)Cl/AlEt₂Cl; x α-TiCl₃/
AlEt₂Cl.

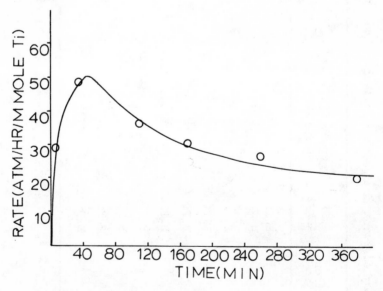

Fig. 9. Propylene polymerization catalyzed by B₃TiCl/Mg(OH)
Cl/AlEt₂Cl at 50°.

324

LITERATURE CITED

(1) P. Cossee, *J. Catalysis* *3*, 80, 90 (1964).

(2) E.J. Arlman, *Proc. Intern. Congr. Catalysis 3rd 2*, 957 (1964).

(3) U.S. Patent 3400110, Solvay & Cie (1963).

(4) British Patents 1140649 and 1137830, Solvay & Cie (1966).

(5) D.G.H. Ballard, *Advances in Catalysis 23*, 293 (1973).

(6) F.J. Karol, G.L. Karapinka, C. Wu, A.W. Dow, R.N. Johnson, and W.L. Carrick, *J. Polymer Sci.*, A-1 *10*, 2621 (1972).

(7) W.L. Carrick, R.J. Turbelt, F.J. Karol, G.L. Karapinka, A.S. Fox, and R.N. Johnson, *J. Polymer Sci.* A-1 *10*, 2609 (1972).

(8) F.K. Karol, G.L. Brown, and J.M. Davison, *J. Polymer Sci.* A-1 *11*, 413 (1973).

(9) K. Soga, S. Katano, Y. Akimoto, and T. Kagiya, *Polymer J*, *5*, 128 (1973).

(10) H.P. Boehm, *Advan. Catal. 16*, 179 (1966).

(11) F.H. Hambleton, J.A. Hockey, and J.A.G. Taylor, *Trans. Faraday Soc.* *6*, 794, 801 (1966).

(12) J.B. Peri, and A.L. Hensley, Jr., *J. Phys. Chem. 72*, 2926 (1968).

(13) J.C.W. Chien, *J. Amer. Chem. Soc. 93*, 4675 (1971).

(14) J.C.W. Chien, and C.R. Boss, *J. Amer. Chem. Soc.*, *83*, 3767 (1961).

(15) D.W. Pratt, Ph.D. Thesis, University of California, Berkeley, Calif. 1967.

(16) R.B. Johannesen, G.A. Candela, and T. Tsang, *J. Chem. Phys. 48*, 5544 (1968).

(17) J.H. DeBoer, and J.M. Vleeskeus, *Proc. Koninkl. Ned Akad. Wetenschap. B60*, 44 (1957).

(18) J.C.W. Chien, *J. Amer. Chem. Soc. 81*, 86 (1959).

(19) E.H. Adema, H.J.M. Bartelink, and J. Smidt, *Rec. Trav. Chim. Pays-Bas*, *80*, 173 (1961).

(20) A. Beermann, and H. Bestian, *Angew. Chem.*, *71*, 618 (1959).

(21) A. Beermann, and K. Clauss, *Angew. Chem. 71*, 627 (1959).

(22) H. deVries, *Rec. Trav. Chim. Pays-Bas*, *80*, 866 (1961).

(23) J.C.W. Chien, *J. Catalysis 23*, 71 (1971).

(24) G. Natta, *Makromol. Chem*, *35*, 94 (1960).

(25) G. Natta, I. Pasquon, and E. Giachetti, *Angew. Chem. 69*, 213 (1957); *Makromol. Chem*, *24*, 258 (1957); *Chimie Ind.* (Milan) *39*, 993 (1957).

(26) G. Natta, I. Pasquon, E. Giochetti, and G. Pajaro, *Chimie Ind.* (Milan) *40*, 267 (1958).

(27) J.C.W. Chien, *J. Polymer Sci. A-1*, 1839 (1963).

(28) J.C.W. Chien, *J. Polymer Sci. A-1*, 425 (1963).

Stereospecific Polymerization of Dioelfins by h³-Allylic Coordination Complexes

PH. TEYSSIÉ, M. JULÉMONT, J.M. THOMASSIN,
E. WALCKIERS, and R. WARIN

*Laboratory of Macromolecular Chemistry
and Organic Catalysis
Université de Liège, Sart Tilman
4000 Liège, Belgium*

More than fifteen years have passed since the first
stereospecific polymerizations of diolefins were achieved
with Ziegler-Natta catalysts. Since then more and more
efficient systems have been discovered which allow greater
structural control, milder reaction conditions, higher reaction rates, and smaller amounts of catalysts. Our mechanistic knowledge of these reactions has also progressed accordingly, and we begin to understand a number of the basic
factors by which they are governed. However these studies
have also revealed an increasingly complex picture of the
behaviour of these catalysts, and we are still far from
having a detailed and satisfying picture of their mode of
action. In fact, it becomes evident that we have under-
estimated the refinement of these controls, and in particular
the determinant role of rather small modifications in the
overall geometry of the complexes in the reaction mixtures.
The present review, which does not intend to be exhaus-
tive, will try to present the most significant advances made
in the field as well as some of the key problems that they
still raise.

1) *FUNDAMENTALS OF THE COORDINATED SPECIFIC POLYMERIZATION
OF DIOLEFINS BY ZIEGLER-NATTA CATALYSTS.*

1.1) Formation of the Active Centers

The work of the past decade on Ziegler-Natta catalysts
(consisting usually of a transition metal derivative and a
metal alkyl or hydride, combined eventually with additional

ligands), established that the growing polymer chain is attached to a metal atom, and the transition metal is determinant in controlling the stereospecific growth reaction (apparently this role can also be played by lithium in coordinated-anionic polymerization). Furthermore, owing to the monomer structure, the catalytic entity involves a metal-allyl type of bonding with the chain. Most of the data presently available on transition metal-allyl systems indicate that the π-electrons of the allyl group are delocalized, and that most of the time one deals with a stabilized π- or h^3-allylic structure (we will adopt Cotton's notation (1), which is unambiguous and versatile). In the case of diene polymerization, the active center may accordingly be represented by:

$$XnM_TY.(L)_z \xrightarrow{\;C_4H_6\;} H-C \overset{CH_2}{\underset{CH_2Y}{\overset{CH}{\diagdown}}} M_T \overset{X_n(L)_z}{\diagdown}_{1-2} \quad \longleftarrow \quad C_4H_6Y \underset{C_4H_6}{\overset{}{\diagdown}} M \overset{X_n(L)}{\diagdown} \underset{C_4H_6}{\diagdown}$$

where Y = R, H, X; X = simple anion; and L = Lewis base, or AlR_xX_{3-x} in the case of a bimetallic Ziegler catalyst (where this aluminum derivative plays the role of a ligand able to influence the rate and eventually the stereospecificity of the reaction).

In support of these ideas, many h^3-allyl complexes of transition metals have been isolated in a pure state(2,3) and found to be reasonably active catalysts for stereospecific diolefin addition reactions. Consequently they represent good models in studying the corresponding polymerization as well as open-chain- and cyclo-oligomerization catalytic reactions(4).

Interestingly, it has been shown that the rate of the process depends directly on the electronic density in specific points of the coordination sphere. For instance in a series of h^3-allylnickel complexes, all of which yield pure *cis*-1,4-polybutadiene, the rate may be varied by more than a thousand-fold simply by increasing the electron-withdrawing character of the counter-anion(5). It is also possible to synthesize these basic monometallic complexes which have activities comparable(6) to the standard bimetallic Ziegler-Natta systems (where the aluminum derivative probably plays the role of a ligand controlling the right electronic distribution).

1.2) The Propagation Mechanism

The polymerization process may now be described further in terms of two very different types of mechanisms. The first one is the *cis-rearrangement reaction* proposed by P. Cossee some 15 years ago to explain the 1,4-polymerization of butadiene by titanium chloride catalysts(7). It involves essentially a π-coordination of the monomer (mono- or biden-tate depending on the number of positions available) on the σ- or h^1-allyl form of the complex, followed by an electronic rearrangement involving some migration of the h^1-bonded group to the coordinated monomer (Fig. 1). It has also been invoked in mechanistic interpretations of lithium-initiated diolefins polymerizations(7).

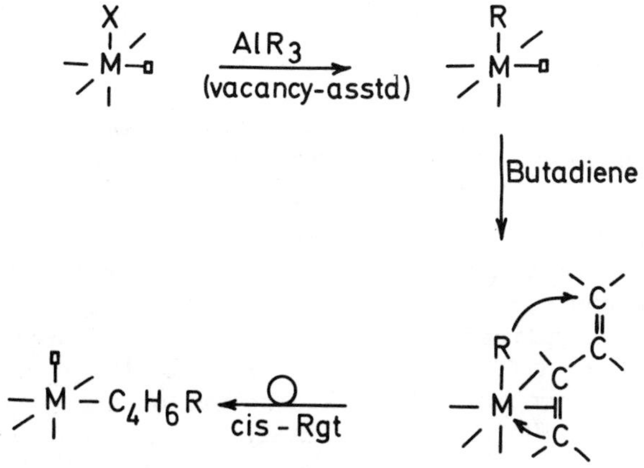

Fig. 1. Cis-rearrangement mechanism in coordination polymerization.

On the other hand, several authors(8,9,10,11) have considered a completely different type of electronic rearrangement, closely to the *allylic transposition* invoked i.e., in the addition of metal-allyl derivatives to ketones(12). In the case of palladium-catalyzed 1,2-addition reactions(10), the new carbon-carbon bond has been shown indeed to be formed outside the coordination sphere. It implies first (Fig. 2) the reversible coordination of monomer with simultaneous conversion of the complex into the h^1-form, involving a σ-bond between the metal and the *least substituted* carbon atom of the allyl group. Furthermore, the bonding of the h^1-allyl group (C^3 atom) to the free C^4-carbon of the coordinated diene proceeds probably through a concerted process.

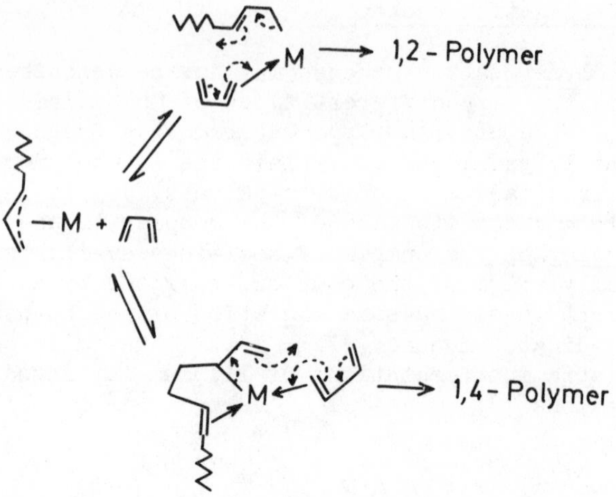

*Fig. 2. Allylic transposition mechanism
(outer sphere pericyclic rearrangement)
in coordination polymerization. Courtesy
of J. Amer. Chem. Soc., 94, 7731 (1972).*

This mechanism has been tentatively extended to 1,4-polymerization by other transition metals, like nickel, even though the stereochemistry involved is completely different(12). However, the application of this scheme to 1,4-polymerization raises several difficulties. It gives no interpretation of the *cis-trans* controls discussed below; on the other hand, it would imply(10) several unlikely situations, such as coordination of the diene by the most substituted double bond (13), and σ-bonding of the h^1-allyl group to the metal by the carbon atom carrying the growing chain. Anyhow, both mechanisms imply a transitory h^1-allyl complex carrying a π-coordinated monomer molecule; this model will be adopted for further discussion as it is well-substantiated by the general chemistry of the allyl complexes.

Further investigations should determine if *cis*-rearrangement predominates in 1,4-addition (butadiene 1,4-polymerization by Ti, Ni, and Li catalysts), while allylic transposition is the main process in the formation of branched addition products (butadiene 1,2-polymerization by Pd or Li-OR catalysts, addition of allyl-Grignard reagents to ketones, piperylene cyclodimerization by nickel complexes).

2) *THE STEREOREGULATION OF THE PROPAGATION PROCESSES IN TRANSITION-METALS CATALYZED POLYMERIZATION.*

2.1) Control of Structural Isomerism (1,2/1,4)

It is by now well established that the structural isomerism of the polydienes obtained by coordination catalysis on transition metal complexes depends essentially on the nature of the metal involved, although in some cases specific ligands can influence it to some extent. For example, 1,2-polybutadiene can be obtained in the presence of palladium, chromium and molybdenum derivatives, while 1,4-polymers are obtained in the presence of titanium, cobalt, nickel, and rhodium complexes (see i.e., (4)). In both cases steric purities higher than 95% are easily obtained, reaching often 98 to 99%.

The lack of experimental methods allowing a progressive variation of this structural isomerism (from pure 1,2 to pure 1,4) by a systematic modification of the catalyst structure, prevented up to now a thorough investigation of this change in microstructures. A combination of electronic and geometrical factors is probably involved. Till now however, the only simple mechanistic proposal remains that of Arlman(18), based on the respective distances between the C_2 or C_4 atoms of the coordinated diolefin and the α-CH_2 group of the h^1-allyl-metal undergoing the rearrangement. This distinction might eventually be coupled with a change in mechanism from *cis*-rearrangement to allylic transposition, or alternatively to a change in the geometry of the complex undergoing this allylic transposition(10). Obviously, more experiments, i.e., with dienes substituted in specific positions, are required to solve these problems.

2.2) Control of the Geometrical Isomerism (cis 1,4/trans 1,4).

2.2.1) *Influence of the Ligands.* As already mentioned, bis-(h^3-allylnickel trifluoroacetate), or $(ANiTFA)_2$, promotes the rapid formation of a very high *cis*-1,4-polybutadiene(5). However, in the presence of different additional ligands, the geometrical isomerism of the polymer obtained changes drastically as indicated in Table 1., while the structural isomerism remains unchanged (99% 1,4). This change from *cis* to *trans* isomerism, common to several catalysts, has been tentatively interpreted by different theories.

TABLE 1.

Control of geometrical isomerism by additional ligands in the
1,4-butadiene polymerization by (ANiTFA)$_2$.

L	[L] / [Ni]	% *cis*	% *trans*
---	---	98	1
P(OR)$_3$	1	---	99
C$_2$H$_5$OH	1	---	99
CF$_3$COOH	1	50	49
C$_6$H$_{6-n}$X$_n$	< 500	48	50

The first one, proposed by Otsuka in 1965(14), is based
on the existence of two different isomers of the h^3-allyl
bonded chain: the *anti* intermediate, promoting the formation
of a *cis* double bond in the chain upon insertion of the next
monomeric unit and reformation of a new h^3-allyl group; and
the *syn* intermediate promoting the formation of a *trans* unit
(see Fig. 3). This interpretation has two strong drawbacks.
First, several NMR studies(15,16,17) have shown that various
h^3-allyl catalysts, giving *cis* or *trans* or mixed polybutadienes, are always in the most probable *syn* form. This result
cannot exclude of course the possibility that a transient
anti form might lead to *cis* polymerization. There is, however,
no reason to believe that the *anti* form has much higher
activity than the *syn* one. Furthermore, such a control might
imply a reaction between two π-bonded entities (h^3-allyl
chain and π-butadiene), which is not too likely in the
chemistry of allyl-metal complexes.

Another interpretation(7,18), proposed by Cossee is
simply based on a monodentate-*trans* or bidentate-*cis* coordination of the diolefin which leads respectively to the formation of *trans* or *cis* configurations in the polymer (see Fig.
4). This hypothesis is strongly supported by the fact that
α-TiCl$_3$, which offers only one coordination vacancy at the
active center, promotes the formation of *trans*-1,4-polymeri-
zation, whereas β-TiCl$_3$ which offers more vacancies favours
the formation of a mixture of homo-*cis* and homo-*trans* polymers.

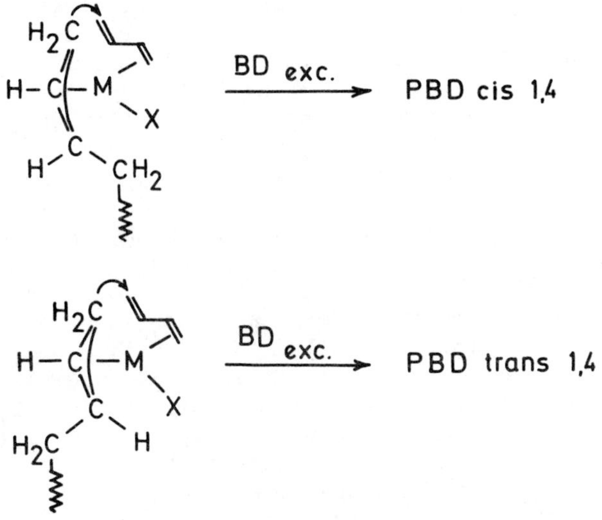

Fig. 3. *Control of the cis or trans config-uration of the h^3-allyl group isomerism in 1,4-diolefins polymerizations.*

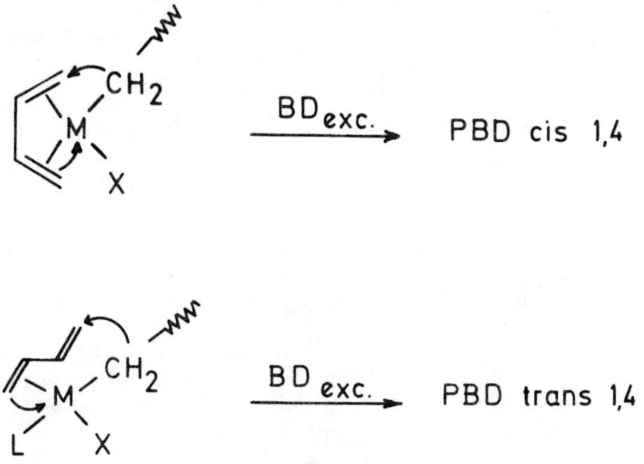

Fig. 4. *Control of the cis or trans config-urations by the coordinated monomer conforma-tion in 1,4-diolefins polymerizations.*

This mechanism also can explain the influence of strong
basic ligands (phosphites, alcohols) which converts the *cis*-
catalyst into a *trans*-catalyst. These ligands are known to
dissociate the binuclear complex and to occupy one coordination position on the mononuclear form. In this way, they
prevent the bidentate coordination of the monomer on the
mononuclear complex(19), which is necessary to generate the
1,4-*cis* polymer(20).

This very straightforward and appealing mechanism, which
is also invoked in several interpretations of lithium-induced
anionic polymerizations, presents, however, one serious problem: indeed if inbetween every insertion step the h^1-allyl
chain returns to the *syn* h^3-allyl form (which is the thermo-
dynamically favoured one for transition metals), the *cis*-
configuration formed in the h^1-allyl chain will be lost, and
the *syn* h^3-allyl bond will most probably regenerate a *trans*
configuration after insertion of the next monomer. Consequentl;
one is forced to postulate that the h^1-allyl chain must have a
very high reactivity and is able to insert many monomeric units
in this form before returning to the h^3-allyl form, and that
the latter is the "dormant" state of the active complex.

With this mechanism, one has a plausible explanation for
the control of *cis* or *trans* isomerism, as well as of the di-
isomeric specificities discussed in the next section (3.3).

Finally, an elegant alternative explanation has been proposed more recently by Furukawa(21) suggesting a control of
the *cis* isomerism by the coordination (or "backbiting") of
the first free double-bond in the chain to the nickel atom,
whatever the *syn* or *anti* structure of the h^3-allyl group.
However, this interesting proposal, which will have to be
ascertained by experimental data, raises also several serious
problems in terms of intromolecular movements and electronic
charge displacements.

2.2.2) *Influence of the Counter-anions.* Using a series of
bis-(h^3-allylnickel-X)complexes, one can also obtain pure
(99%) 1,4-polybutadienes, whose geometrical isomerism ranges
from 98% *cis* (X = CF_3COO) to 99% *trans* (X = I), with a monotonic variation through Cl and Br (6). This control seems
to depend more on the electronegativity of the counter-anion
than on its bulkiness, as indicated by the formation of very
high *cis*-1,4-polybutadiene from h^3-allylnickel-1,3,5-trinitro-
phenate(5) (picrate salt). This kind of dependence might of
course imply a more or less easy dissociation of the initially
binuclear h^3-allyl complexes, leading or not to a mononuclear
complex capable of accomodating the monomer in a bidentate
coordination (i.e., able to give the *cis*-1,4-isomer, see 2.2.1).

However, several kinetics and structural studies have shown that the formation of both pure *cis* (20) and pure *trans* (22, 23) polybutadienes imply most probably a transient mononuclear species; accordingly it seems more likely that it is the electron withdrawing character of the counter-anion which influences the geometry of the active complex and favours the coordination of the double bond of the second monomer. It could also change the M-X distance and consequently the effective steric hindrance of the anion to this bidentate coordination.

Anyhow, the profound influence of the anion electronegativity on the overall structure of the complex, and in particular of the h^3-allyl group, has been confirmed by NMR spectroscopy(17). More electronegative anions induce a higher dissymmetry of the allyl group (higher bond order for C_2-C_3 than for C_1-C_2, although the group remains h^3-bonded), promoting a resultant twisting of the first methylene group of the chain.

In conclusion, this important question of the *cis/trans* isomerism control is still far from being settled, although the mode of the monomer coordination in a *cis*-rearrangement mechanism seems to offer an attractive straightforward explanation of the results obtained. Again, more experimental results are needed to further our understanding of this specificity.

2.3) Control of Specific Binary Isomerisms: the Case of Equibinary Polydienes

2.3.1) *Overall characteristics of equibinary polymerizations.* In the course of a study of butadiene polymerization by the very active complex, bis(h^3-allylnickel-trifluoroacetate), it was discovered a few years ago(29) that the addition of specific ligands could lead to the formation of a polymer containing exactly equal amounts of *cis* and *trans* isomers. It was later shown that this type of behaviour is quite general in coordination polymerization, and the name "equibinary polydienes" has been coined for these polymers containing equal amounts of two different isomeric units. Typical examples in addition to the equibinary (*cis* 1,4-*trans* 1,4) polybutadiene described above are the equibinary (1,4-1,2) polybutadiene(25), (*cis* 1,4-3,4) polyisoprene(26), and (1,2-3,4) polyisoprene(26), obtained in the presence of various metals like nickel, molybdenum and cobalt. It soon became evident that this phenomenon exhibited all the characteristic features of a specific coordination competitive reaction, as summarized in Fig. 5.

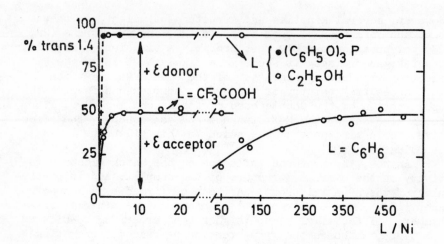

*Fig. 5. Coordination equilibria in equibinary 1,4-polybuta-
diene formation. (Courtesy of J. Polymer Sci. A1, 987
(1970)).*

(1) This equibinary composition is reached asymptotic-
ally when adding increasing amounts of ligand (i.e., olefinic,
aromatic or chlorinated hydrocarbons, or trifluoroacetic acid).
Furthermore, the 50/50 composition is obtained with more or
less important proportions of ligand, depending on the nature
of this ligand; in other words, a greater curvature of the
plots in Fig. 5 might reflect a higher relative stability
(K_f) of the new active entity L-Allyl Ni (-BD).

(2) The phenomenon is reversible as expected for a
simple coordination control: elimination of the ligand
(i.e., by evaporation) or addition of stronger ligands
either donor or acceptor), affects the stereospecificity.

(3) Finally, it should be stressed that the complex is
highly specific. The amount of a third isomer remains very
low; i.e., in the case of (1,4 *cis*-1,4 *trans*) polybutadiene,
the content of 1,2 units is usually lower than 1%. On the
other hand, the equibinary composition can be obtained in a
very broad range of concentrations and temperatures (from
-10 to +70°).

These data suggest that the formation of the equibinary
polymers depends on a new, different, and specific catalytic
center, as shown by the fractionation of a 1,4-polybutadiene
having an intermediate geometric composition (Table 2).
(This sample was obtained by adding a limited amount of
aromatic ligands to the complex generating the *cis*-polymer).

TABLE 2.

Fractionation of a 1,4 polybutadiene of intermediate composition.

Product	% weight	% *cis* 1,4
Total crude	100	66
Insol. in C_6H_6/CH_3OH	70	75
Sol. in C_6H_6/CH_3OH	30	48

The results show clearly that there are two different catalytic centers in equilibrium, the first one yielding a high *cis*, and the other one the equibinary polybutadiene. The same fractionation of an equibinary 1,4-polybutadiene does not change its 50/50 composition(27).

It should be emphasized that these *cis/trans* isomerisms are directly and irreversibly controlled by the catalyst during the polymerization reactions. In other words, a pre-formed polybutadiene polymer cannot be isomerized by the catalyst.

2.3.2) *Structure of the active center.* NMR studies have revealed that the complex carrying the growing chain in the equibinary 1,4-butadiene polymerization ((ANiTFA)$_2$ in benzene) is predominently in the form of a binuclear *syn*-h^3-allylnickel complex. Obviously, these results do not prove the actual structure of the complexes in the active state, and could well represent a "dormant" situation. (That could be corroborated by a definite increase of resolution in the spectra when the reaction is stopped(17), by lowering the temperature of exhausting the monomer). In particular, it is most probable that the reaction proceeds through a h^1-allyl type of bonding of the polymer chain in the insertion step.

The binuclear structure of the active catalytic center is supported by the results gathered in Fig. 6. Comparison with Fig. 5 showed that under conditions specific for equibinary polymerization, a decrease in catalyst concentration causes a return to the specificity characteristic of the complex before addition of the solvent-ligand (this being eventually the monomer itself), although the same amounts of this

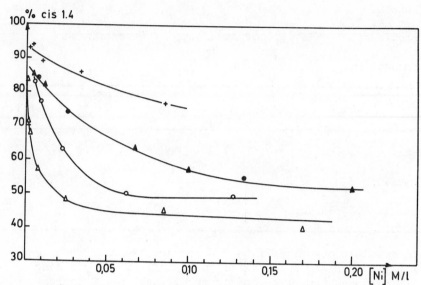

Fig. 6. Competition between cis 1,4- and equibinary 1,4-orientations in function of catalyst concentration, in butadiene polymerizations by (ANiTFA)$_2$. + Polymerization in heptane, 40°C; ● Bulk polymerization, 5°C; ▲ Bulk polymerization, 40°C; ○ Polymerization in benzene, 5°C; △ Polymerization in benzene, 40°C; [BD] = 2.5 M when polymerized in solution.

solvent-ligand is still present. Even in *n*-heptane, where the catalyst gives a high *cis*-content, an increase of this catalyst concentration leads to a definite increase in the *trans* content.

These key-experimental results can be easily explained (20) if one assumes that the equibinary polymer is formed on a binuclear complex, which dissociates upon dilution to give a mononuclear species; the latter is responsible for the formation of pure *cis*-polybutadiene because this complex is capable of accomodating the monomer in the bidentate conformation. Alternatively, the formation of the *cis*-isomer in the equibinary polymer might imply a temporary rupture of one of the bridging bond of the binuclear complexes, to liberate the necessary additional coordination position; the severed bridging bond being restored after monomer insertion to the binuclear structure. The role of the ligand-solvent is probably to stabilize the binuclear form (i.e., by coordination in an apical position), against the different competing coordination equilibria (monomer, spontaneous dissociation, ...).

2.3.3) *Distribution of the equibinary isomeric units in the chain: a new type of control of the polymer structure.* To account for this 50/50 specific composition, several hypotheses can be considered a priori(27).

(1) The formation of equal weights of *cis*-1,4- and *trans*-1,4-polymeric chains, either independent (homopolymers) or linked together in a stereoblock polymer. These two possibilities have no precedence in coordination polymerization, and can be ruled out (at least for the case of long stereosequences) by the fact that the polymer cannot be separated into fractions rich in *cis* or *trans* units. The polymer also has none of the physical properties characteristic of the pure *cis*- or *trans*-polybutadiene.

(2) Either a perfectly random or a perfectly alternating distribution of the *cis* and *trans* isomeric units along the chain. This problem has been approached by two different methods. Equibinary polymers of structural isomers such as equibinary poly(1,2-1,4)butadiene can be ozonolyzed, and analysis of the ozonolysis products permits a determination of the distribution. This method is however not applicable to equibinary polymers involving two geometrical isomers. For these polymers, one can determine the distribution with high resolution ^{1}H NMR with spin decoupling(28).

The results obtained are quite astonishing, and demonstrate again that structural controls by the coordination complexes are extremely sensitive to the reaction conditions. Poly (*cis*-1,4-1,2) butadiene(29) prepared in the conventional manner has a random distribution of the isomer units in the chain as shown by ozonolysis and NMR. The equibinary poly (*cis*-1,4-3,4) isoprene has also been studied; the ozonolysis products indicate that this polymer contains large blocks of head-tail, head-head, and tail-tail 1,4-polyisoprene units (30).

A detailed study of poly (*cis* 1,4-*trans* 1,4) butadiene formation in the presence of (ANiTFA)$_2$, revealed a still much more complex behaviour(31). Polymerization in hydrocarbon solvents (heptane + CF$_3$COOH, benzene) gave a product which is perfectly random in its distribution of the *cis* and *trans* isomers according to NMR. This distribution fits a Bernoullian type statistics, and is similar to that obtained by isomerizing a pure *cis*-1,4-polybutadiene to a 50/50 composition by usual chemical techniques(32) (Fig. 7A). In other solvents like dichloromethane, the equibinary polybutadiene obtained displays a distinct preference ($\sim$65%) for alternating placement of *cis* and *trans* units (Fig. 7B) following a first-order Markov statistics.

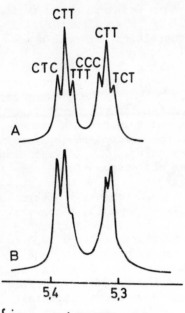

Fig. 7. High-resolution ^1H NMR spectra (olefinic protons) of random (A), and highly alternating (B) equibinary 1-4-polybutadienes.

δ in ppm (HMDS = 0)

In conclusion, the catalyst system described here appears capable of exerting certain degree of controlling the distribution of the isomeric units along the chain, without affecting the equibinary (50/50) composition(33). This control depends essentially on the temperature and the nature of the solvents, as may be expected for a coordination equilibrium between metal and solvent (ligand). This represents a new type of specific catalytic control in the coordination polymerization of diolefins.

Investigation of intermediate compositions, i.e. 75% *cis*-1,4/25%-*trans*-1,4, have indicated distributions of the Coleman-Fox type suggesting again the presence of two independent centers in equilibrium, one promoting the formation of equibinary polymer and the other one that of a stereoblock polybutadiene (predominantly *cis*-1,4), in agreement with the fractionation experiments reported above.

The differences found in the isomeric units distribution for polymers having the same 50% *cis* - 50% *trans* composition, raise an interesting mechanistic problem. Assuming that this type of control arises from a binuclear catalytic center, stabilized by different ligands or solvents, a random placement of pure equibinary 1,4 structure over a broad range of conditions implies probably a mutual control between two growing centers bound in the same specific complex. This

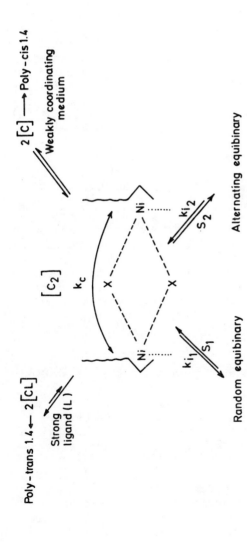

Fig. 8. Tentative scheme of the different structural controls involved in specific 1,4-butadiene polymerizations.

hypothesis fits with the fact that there is indeed one growing chain per nickel atom (see below).

A very tentative interpretation of the alternating distribution control is a modification of the relative kinetics of the monomer insertion by reaction temperature and solvent (versus some geometrical rearrangement in the binuclear complex). In other words, there would be a thermodynamic control of the equibinary phenomenon (binuclear complex stabilization), and a kinetic control of the degree of alternation (relative rate of exchange of the *cis*- and *trans*-controlling positions versus the rate of monomer insertion).

In summary, although it is still too early to propose a detailed mechanism for these controls, one could suggest tentatively the following scheme (Fig. 8) for the 1,4-polymerization of butadiene by bis(h^3-allylnickel-trifluoroacetate). It might be also worthwhile to stress that equibinary polymerization may be a general phenomenon in coordination catalysis as suggested by several recent results in copolymerization of olefins(34), in catalytic cyclopropanation of olefins(35,36) and in coordination isomerization of allylamido derivatives (37,38) (yielding in both cases 50/50 mixtures of *cis* and *trans* isomers in limiting conditions).

3)   *COURSE OF THE DIOLEFINS COORDINATION POLYMERIZATION: THE DIFFERENT REACTION STEPS.*

It may be useful to conclude this review by a summary of the most characteristic kinetic features of these polymerization reactions initiated by transition metal complexes, based on the results obtained in detailed studies on the behaviour of two representative catalysts: bis(h^3-allylnickel iodide) or (ANiI)$_2$, and bis(h^3-allylnickel trifluoroacetate), or (ANi-TFA)$_2$.

3.1) Initiation Step.

Two important points emerge from the different studies already published.

(1)   The allyl group initiates the polymeric chain and becomes the terminus of this chain, formed by successive insertions of monomer molecules. The reaction scheme given below has been ascertained by g.l.c.(39), infrared spectrometry(39), and NMR measurements(17). The structure of the products obtained by using a h^3-crotyl complex(17) suggests that a *cis*-rearrangement mechanism is preferred over an outer-sphere electrocyclic rearrangement of the Powell's type (this latter mechanism would require the unlikely pre-

sence of the chain on the carbon atom directly σ-bonded to the metal in the reactive h^1-allyl form).

Initiation $\downarrow$ 1 $C_4H_2D_4$

Propagation $\downarrow$ n $C_4H_2D_4$

(2) The initiating complex is rather rapidly consumed even at low monomer/catalyst (M/C) ratios(39,40). This implies that every metal atom can be active even if the catalytic centers remain under the form of a binuclear complex. In fact the propagating center differs from the original complex only in the substitution of the C^3 in the allyl group by the polymer chain. With substituted dienes, the initiation rate may be higher while the propagation rate is lower(41,42) than with butadiene.

3.2) Propagation Step.

Relative initiation and propagation rates have been determined by NMR measurements on the polymerizing mixture

(39,40). From the ratios of the CH_2^c and CH_2^t resonance peaks, one finds that k_p is certainly greater than k_i, but both have the same order of magnitude. This is expected from complexes having very similar structures. The reaction rate is usually first order in monomer concentration, in agreement with the above coordination scheme for a binuclear complex. The order in catalyst concentration depends on several factors. With $(ANiI)_2$, it varies from 0.5 to 1 (43,22,23,44) with decreasing catalyst concentration suggesting that under these conditions *trans*-1,4 polymerization takes place on a mononuclear species formed in a rate-determining dissociation step. With $(ANiTFA)_2$ under conditions yielding the equibinary polymer, i.e., aromatic solvents, the kinetics is first order in catalyst concentration(45); this corresponds to two chains growing on a binuclear catalytic complex and every nickel atom being active.

$\Delta S^{\ddagger}$ and $\Delta H^{\ddagger}$ have the usual values for this type of polymerization, i.e. around −20 e.u. (23) and 10 to 15 Kcal/ Mole(23,43,45), respectively. It has also been shown in several cases that the overall rate of these processes usually increase with the polarity of the solvents(22,46).

3.3) Transfer Reactions.

The fact that every nickel atom is active and initiates one growing chain by these catalysts suggests a living system. Behaviors characteristic of a living polymerization have been indeed observed; one such example is shown by seeding-resumption experiments(39). However, at higher M/C ratios, the molecular weights do not increase anymore, indicating the interference of transfer reactions(22,39). Depending on the catalyst structure, they are more or less influenced by solvent; for example o-dichlorobenzene can greatly decrease these chain transfers(39).

A mechanistic proposal based on a hydride shift(47) has been confirmed by NMR spectra, indicating(39) the presence of crotyl complexes at the end of polymerizations initiated by $(ANiTFA)_2$. Transfer reactions have also been observed when the structure of the active site is altered during the course of a polymerization to synthesize block-copolymers or -stereoisomers(46).

In particular, the addition of styrene to living equibinary polybutadiene resulted in the formation of two different homopolymers easily separated by fractionation. Similarly the addition of butadiene containing $P(OR)_3$ to the same living polymer in order to make a stereoblock containing one poly*cis* and one poly*trans* sequence resulted in the formation of the two independent pure polymeric isomers.

4) CONCLUSIONS

Our understanding of the diolefin coordination polymerization has clearly improved during the past decade. The basic structure of the catalyst has been elucidated by studies on well-defined allyl-complexes, and a detailed determination of the polymers microstructure has lead to a better knowledge of the mechanistic pathways.

There is no doubt however that much remains to be done in two particular directions. From a preparative point of view, the synthesis of equibinary polymers with a perfectly alternating isomers distribution is a worthwhile goal. So is the preparation of living high molecular weight polybutadienes and their block copolymers. From a mechanistical point of view, it would be most interesting on the one hand to clarify the nature of the electronic rearrangements involved (specially in terms of the rules governing the concerted reactions). On the other hand, it is becoming obvious that these elaborated stereocontrols depend on the overall geometry of the active complexes in the reaction medium. It would be most helpful to accumulate experimental informations about this actual structure in order to devise still better control of the polymers stereoregularities. It is highly probable that this knowledge can be obtained only through the simultaneous application of several physical methods such as magnetic measurements (susceptibility, EPR, NMR) and optical spectroscopy.

Even without taking into account possible unexpected new breakthroughs in the field, there are still a lot of exciting challenges in this fascinating area of research where, by very different routes, the chemists have attained a degree of activity and specificity nearly comparable to those encountered in biological systems.

LITERATURE CITED

(1) F.A. Cotton, *J. Am. Chem. Soc.*, *90*, 6230 (1968).
(2) F. Dawans, J.C. Maréchal, and Ph. Teyssié, *J. Organometal. Chem.*, *21*, 259 (1970).
(3) F. Dawans and Ph. Teyssié, *J. Polymer Sci.* *B7*, 111 (1969).
(4) F. Dawans and Ph. Teyssié, *Ind. Eng. Chem. Prod. Res. Develop.*, *10*, 263 (1971).
(5) J.P. Durand, F. Dawans, and Ph. Teyssié, *J. Polymer Sci.* *A1*, 979 (1970); see also ref. (3).
(6) J.P. Durand, F. Dawans and Ph. Teyssié, *J. Polymer Sci.* *B6*, 760 (1968).

(7) P. Cossee, Stereochemistry of Macromolecules, Vol. I, p. 145, M. Dekker, New York 1967; M. Morton et al., *Macromolecules, 6,* 181 (1973).

(8) Y. Sokolov, and I.Y. Poddubnyi, *Adv. Chem. Ser., 91,* 250 (1969).

(9) Y. Takahashi, S. Sakai, and Y. Ishii, *J. Organometal. Chem., 16,* 177 (1969).

(10) R.P. Hughes, and J. Powell, *J. Am. Chem. Soc., 94,* 7723 (1972).

(11) P. Heimbach, *Angew. Chem. Int. Ed., 12,* 975 (1973).

(12) Ph. Teyssié, M. Julémont, E. Walckiers, and R. Warin, Encyclopedia of Polymer Science and Technology, in preparation.

(13) C.A. Tolman, *J. Am. Chem. Soc., 96,* 2780 (1974).

(14) S. Otsuka, and M. Kawakami, *Kogyo Kagaku Zasshi, 68,* 874 (1965).

(15) T. Matsumoto, and J. Furukawa, *J. Polymer Sci., B5,* 935 (1967).

(16) M.I. Lobach, *J. Polymer Sci., B9,* 71 (1971); V.A. Kormer, and M.I. Lobach, *ibid., B10,* 177 (1972).

(17) R. Warin, Ph. Teyssié, P. Bourdauducq, and F. Dawans, *J. Polymer Sci., B11,* 177 (1973).

(18) E.J. Arlman, *J. Catalysis, 5,* 178 (1966); see also ref. (7).

(19) G. Natta, U. Giannini, P. Pino, and A. Cassata, *Chim. e Ind., 47,* 524 (1965).

(20) E. Walckiers, M. Julémont, J.M. Thomassin, and Ph. Teyssié, in preparation.

(21) J. Furukawa, Plenary Lecture, IUPAC Symposium on Macromolecules, Madrid 1974.

(22) G. Henrici-Olivé, S. Olivé, and J. Schmidt, *J. Organometal. Chem., 39,* 201 (1972).

(23) J. F. Harrod, and L.R. Wallace, *Macromolecules, 2,* 449 (1969); *ibid., 5,* 682 (1972).

(24) Ph. Teyssié, F. Dawans, and J.P. Durand, *J. Polymer Sci., C22,* 221 (1968); J. P. Durand, and Ph. Teyssié, *J. Polymer Sci., B6,* 299 (1968); J.P. Durand, F. Dawans, and Ph. Teyssié, *J. Polymer Sci., A1,* 987 (1970).

(25) J. Furukawa, K. Haga, E. Kobayashi, Y. Iseda, T. Youshimoto, and K. Sakamoto, *Polymer J., 2,* 371 (1971); J. Furukawa, E. Kobayashi, and T. Kawagoe, *ibid., 5,* 231 (1973).

(26) F. Dawans, and Ph. Teyssié, *Makromolekulare Chem., 109,* 68 (1967).

(27) J.M. Thomassin, E. Walckiers, R. Warin and Ph. Teyssié, *J. Polymer Sci., B11,* 229 (1973).

(28) E.R. Santee, V.D. Mochel, and M. Morton, *J. Polymer Sci., B11,* 453 (1973).

(29) J. Furukawa, K. Haga, E. Kobayashi, Y. Iseda, T. Yoshi-
      moto, and K. Sakamoto, *Polymer J.*, *2*, 371 (1971); J.
      Funkawa, E. Kobayashi, and T. Kawagoe, *Polymer J.*, *5*,
      231 (1973).
(30) M.J. Hackathorn, and M.J. Brock, *Polymer Preprints*, *14*,
      42 (1973).
(31) M. Julémont, E. Walckiers, R. Warin, and Ph. Teyssié,
      Makromolekulare Chem., *175*, 1673 (1974).
(32) M. Berger, and D.J. Buckley, *J. Polymer Sci.*, *A1*, 2945
      (1963).
(33) M. Julémont, E. Walckiers, R. Warin, and Ph. Teyssié, in
      preparation.
(34) F. Dawans and Ph. Teyssié, *Europ. Polym. J.*, *5*, 541
      (1969).
(35) R. Paulissen, A.J. Hubert, and Ph. Teyssié, *Tetrahedron
      Letters*, 1465 (1972).
(36) R. Paulissen, Ph.D. Thesis, Univ. of Liège, 1974.
(37) A.J. Hubert, Ph. Moniotte, G. Goebbels, R. Warin, and
      Ph. Teyssié, *J. Chem. Soc.*, Perkin II, 1955 (1973).
(38) A. Feron, Lic. Thesis, Univ. of Liège, 1974.
(39) E. Walckiers, J.M. Thomassin, R. Warin, and Ph. Teyssié,
      J. Polym. Sci., *A1*, in press (1975).
(40) M.I. Lobach, V.A. Kormer, I.Yu. Tsereteli, G.P. Kondra-
      tenkov, B.D. Babitskii, and V.I. Klepilova, *J. Polymer
      Sci.*, *B9*, 71 (1971).
(41) V.I. Klepikova, V.A. Vasiliev, G.P. Kondratenkov, and
      M.I. Lobach, *Dokl. Chem.*, *211*, 641 (1973).
(42) T.B. Soboleva, V.A. Yakovlev, E.I. Tinyakova, and B.A.
      Dolgoplosk, *Dokl. Chem.*, *212*, 810 (1973).
(43) A.M. Lazutkin, V.A. Vashkevich, S.S. Medvedev, and V.N.
      Vasil'eva, *Dokl. Phys. Chem.*, *175*, 583 (1967).
(44) A.I. Lazutkina, L.Y. Alt, T.L. Matveeva, A.M. Lazutkin,
      and Y.I. Ermakov, *Kin. i Katal.*, *12*, 1162 (1971).
(45) J.C. Maréchal, F. Dawans, and Ph. Teyssié, *J. Polymer
      Sci.*, *A1*, 2003 (1970).
(46) J.M. Thomassin, Ph.D. Thesis, Univ. of Liegé, 1975.
(47) J.F. Harrod and L.R. Wallace, *Macromolecules*, *5*, 685
      (1972).

SUBJECT INDEX

A

Aufbaureaktion, 9, 11
Active Site concentration in
  supported catalysts, 97
  $TiCl_2$, 115
  $TiCl_2/AlEt_2Cl$, 100, 102, 115
  $TiCl_3$ catalysts, 142
  VCl_3/AlR_3, 164
    modified by donor, 194
Active site counting, 74-88, 92-130, 138-146
  by $Al(^{14}CH_3)_2Cl/(C_5H_5)_2TiCl_2$, 75
  by $Al_2(^{14}C_2H_5)_3Cl_3/VOCl_3$, 75
  by CH_3O^3H, 94, 102
  by $i\text{-}C_3H_7O^3H$, 75
  by $n\text{-}C_4H_9O^3H$, 77
  by ^{14}CO, 95, 102, 138
  by $^{14}CO/PH_3$, 95, 102
  by $^{35}SO_2$, 99, 102
  isotope effect, 78
Active species
  of h^3-allylnickel trifluoroacetate, 337
  ionic, 210
    solvent influence, 212
  valence of Ti, 148
Additoni of monomer
  cis, 16, 61
  trans, 16
Adsorption of
  aluminum alkyls, 119
  monomer, 156
$Al(^{14}CH_3)_2Cl$ in active site counting, 75
$Al_2(^{14}C_2H_5)_3Cl_3/VOCl_3$ in active site
  counting, 75

B

Bromination of allylic olefins, 8

Butadiene, polymerization, 6, 327-345
Butene-1, polymerization, 74-88
  rate constants, 84

C

Chain transfer reaction, 149
  to aluminum alkyls, 184
  to monomer, 184
  spontaneous termination, 183
Chirality, 16
CH_3O^3H in active site counting, 94, 102
$i\text{-}C_3H_7O^3H$ in active site counting, 75
$n\text{-}C_4H_9O^3H$ in active site counting, 77
Cis-rearrangement reaction, 329
^{14}Co in active site counting, 95, 102, 138
  with PH_3, 95, 102
$^{14}CO_2$ in active site counting, 95, 102
Conformational effects in stereoelective
  polymerization, 55
$Cr(allyl)_3/SiO_2$, 94, 98, 108, 111, 112
  metal polymer bond, 97
CrO_3/SiO_2
  metal polymer bond, 97
Copolymer
  ethylene/propylene
    sequence distribution, 19
  4-methylheptene-1/ethylene, 39
  racemic α olefin/optically active
    α olefin, 48
Cyclization of
  α,ω-dinitriles, 7
  butadiene, 8
Cyclic ketones, 7

D

Degree of polymerication, 78, 81

349